Management and Industrial Engineering

Series editor

J. Paulo Davim, Aveiro, Portugal

More information about this series at http://www.springer.com/series/11690

J. Paulo Davim
Editor

Curricula for Sustainability in Higher Education

Springer

Editor
J. Paulo Davim
Department of Mechanical Engineering
University of Aveiro
Aveiro
Portugal

ISSN 2365-0532 ISSN 2365-0540 (electronic)
Management and Industrial Engineering
ISBN 978-3-319-56504-0 ISBN 978-3-319-56505-7 (eBook)
DOI 10.1007/978-3-319-56505-7

Library of Congress Control Number: 2017937722

Printed on acid-free paper

This Springer imprint is published by Springer Nature
The registered company is Springer International Publishing AG
The registered company address is: Gewerbestrasse 11, 6330 Cham, Switzerland

Preface

Currently, universities recognize their responsibility to transform learning in order to promote sustainability. It is important that all university agents (administrators, teachers, students, etc.) diffuse for the development of curricula practices for sustainable education based on three interconnected pillars, the environmental, the economic and the social, along with emerging institutional dimension (politics). The key focus of this book is curricula for sustainability in higher education.

The purpose of this book is to present a collection of examples illustrating curricula for sustainability in higher education. The first chapter of the book provides education for sustainable development and its role in the promotion of the sustainable development goals. The second chapter is dedicated to explicit economics (addressing conscious consumption for sustainability). The third chapter described greening networks (mapping sustainability beyond institutional boundaries). The fourth chapter is dedicated to Preschools Teachers' sustainable competencies within New Kosovo (curriculum framework). The fifth chapter contains information on building energy, environment and sustainability linkages in management education in India (an innovative curriculum based approach). Finally, the sixth chapter is dedicated to thinking about sustainability (issues and themes for college students).

The present book can be used as a valuable reference for academics, researchers, educators, managers, engineers, and professionals in sustainability in higher education and related subjects. The interest of scientific in this book is evident for many important institutes of the research and universities. Consequently, it is expected this book will motivate others to undertake research in sustainability in higher education.

The Editor acknowledges Springer for this opportunity and for their qualified and professional support. Finally, I would like to thank all the chapter authors for their availability for this work.

Aveiro, Portugal
May 2017

J. Paulo Davim

Contents

Editor and Contributors

About the Editor

J. Paulo Davim received his Ph.D. in Mechanical Engineering in 1997, M.Sc. degree in Mechanical Engineering (materials and manufacturing processes) in 1991, Graduate degree (5 years) in Mechanical Engineering in 1986, from the University of Porto (FEUP), the Aggregate title (Full Habilitation) from the University of Coimbra in 2005 and a D.Sc. from London Metropolitan University in 2013. He is Eur Ing by FEANI-Brussels and Senior Chartered Engineer by the Portuguese Institution of Engineers with a MBA and Specialist title in Engineering and Industrial Management. Currently, he is a Professor at the Department of Mechanical Engineering of the University of Aveiro, Portugal. He has more than 30 years of teaching and research experience in Manufacturing, Materials and Mechanical Engineering with a special emphasis on Machining and Tribology. He has also interest in Management/Industrial Engineering and Higher Education for Sustainability/Engineering Education. He has received several scientific awards. He has worked as an evaluator of projects for international research agencies as well as an examiner of Ph.D. thesis for many universities. He is the Editor in Chief of several international journals, Guest Editor of journals, books Editor, book Series Editor and Scientific Advisory for many international journals and conferences. Presently, he is an Editorial Board Member of 30 international journals and acts as reviewer for more than 80 prestigious Web of Science journals. In addition, he has also published as a editor (and co-editor) more than 100 books and as a author (and co-author) more than 10 books, 60 chapters, and 400 articles in journals and conferences (more than 200 articles in journals indexed in Web of Science/h-index 38+ and SCOPUS/h-index 46+).

Contributors

Arlinda Beka Faculty of Education, University of Prishtina, Prishtina, Kosovo

Viraja Bhat Department of Energy and Environment, Symbiosis Institute of International Business, Symbiosis International University, Pune, India

Gregory D. Bothun Department of Physics, University of Oregon, Oregon, USA

Andréia Simas Carvalho University of Southern Santa Catarina (Unisul), Santa Catarina, Brazil

José Baltazar Salgueirinho Osório de Andrade Guerra University of Southern Santa Catarina (Unisul), Santa Catarina, Brazil

Jéssica Garcia University of Southern Santa Catarina (Unisul), Santa Catarina, Brazil

Manisha Ketkar Symbiosis School of Banking and Finance, Symbiosis International University, Pune, India

Naveed Nanjee University of Edinburgh, Edinburgh, UK

Yogesh Patil Department of Energy and Environment, Symbiosis Institute of International Business, Symbiosis International University, Pune, India; Symbiosis Centre for Research and Innovation, Symbiosis International University, Pune, India

Sthefanie Aguiar da Silva University of Southern Santa Catarina (Unisul), Santa Catarina, Brazil

Prakash Rao Department of Energy and Environment, Symbiosis Institute of International Business, Symbiosis International University, Pune, India

Madhavi Venkatesan Department of Economics, Bridgewater State University, Bridgewater, MA, USA

Fayyaz Vellani Critical Writing Program, University of Pennsylvania, Philadelphia, PA, USA

Education for Sustainable Development and Its Role in the Promotion of the Sustainable Development Goals

Jéssica Garcia, Sthefanie Aguiar da Silva, Andréia Simas Carvalho and José Baltazar Salgueirinho Osório de Andrade Guerra

Abstract The Sustainable Development Goals emerged as universal goals aiming to promote more inclusive societies, fighting inequalities and recognizing the importance of cooperation. Based on a human rights approach, the idea is to pursue sustainable development along with human development, stressing the importance of each of the 17 goals set by the 2030 Agenda and how interconnected they all are. At the same time, Education for Sustainable Development is pursuing similar goals, including gender equality, health promotion and climate change action. The main purpose of this paper is to understand how Education for Sustainable Development can assist in the achievement of the Sustainable Development Goals. In addition to a literature review, a discussion is conducted from the presentation of each Sustainable Development Goal and how they can be explored through education. As a safe environment, Education Institutions are a platform that should be used for debates, especially the ones related to the challenges society has to face in the future. In this context, Education for Sustainable Development carries important tools to enable the learners not only to be prepared for what their lives and career will require, but also to contribute for a more sustainable development and to take conscious decisions. To address the challenges within the Sustainable Development Goals requires only the willingness of education leaders and stakeholders to adjust their curricula and engage the teaching body towards the change our world demands.

Keywords Sustainable development goals · Education for sustainable development · Sustainability · Climate change

---

J. Garcia · S.A. da Silva · A.S. Carvalho · J.B.S.O. de Andrade Guerra (✉)
University of Southern Santa Catarina (Unisul), Santa Catarina, Brazil
e-mail: baltazar.guerra@unisul.br

J. Garcia
e-mail: jessica.sgarcia@outlook.com

S.A. da Silva
e-mail: sthefanie.sads@hotmail.com

A.S. Carvalho
e-mail: deia.sccarvalho@gmail.com

J.P. Davim (ed.), *Curricula for Sustainability in Higher Education*,
Management and Industrial Engineering, DOI 10.1007/978-3-319-56505-7_1

1 Introduction

The Sustainable Development Goals (SDGs) were conceived at the Rio +20 Summit, held in Brazil in 2012, to be a continuation of the previous Millennium Development Goals (MDGs) agenda. The Rio +20's "The Future We Want" document highlights the need for the Post 2015 Agenda to have objectives geared towards the economic, social and environmental areas, in a balanced and integrated way, and in need of strong partnerships, with the involvement of all stakeholders (United Nations 2012). According to Sachs (2012, p. 2206), "the idea of the SDGs has quickly gained ground because of the growing urgency of sustainable development for the entire world".

In the context of the outcomes of Rio +20, the SDGs idealization process involved more than three years of negotiation (UNSDSN 2016a, b), involving the Secretary-General of the United Nations, Secretary-General's High-Level Panel of eminent persons on the Post-2015 Development Agenda and the United Nations Sustainable Development Solutions Network, organizations created to designate the implementation of the Sustainable Development Goals, as well as consultations with civil society, academia, private sector and other stakeholders (UNSDSN 2016b; Haslegrave 2014; Post-2015; HLP 2016). After the period of negotiations and consultations, UN Member States met to determine their goals and focus on the post 2015 agenda.

Through the letter "Transforming Our World: The 2030 Agenda for Sustainable Development", signed by the UN Member States in August 2015, the Post 2015 Agenda was determined with an action plan for people, planet and prosperity (United Nations 2015). From this letter, 17 goals were agreed, with 169 subsequent targets, to be met by 2030.

The overall focus of the SDGs was earlier defined in "The Future We Want" document, as stated by government representatives from all participating countries:

> poverty eradication, changing unsustainable and promoting sustainable patterns of consumption and production and protecting and managing the natural resource base of economic and social development are the overarching objectives of and essential requirements for sustainable development. We also reaffirm the need to achieve sustainable development by promoting sustained, inclusive and equitable economic growth, creating greater opportunities for all, reducing inequalities, raising basic standards of living, fostering equitable social development and inclusion, and promoting the integrated and sustainable management of natural resources and ecosystems that supports, inter alia, economic, social and human development while facilitating ecosystem conservation, regeneration and restoration and resilience in the face of new and emerging challenges (United Nations 2012, p. 1).

Following these focal points and taking upon the interests and needs of States previously discussed in "Transforming Our World: The 2030 Agenda for Sustainable Development", the objectives presented in Table 1 were agreed upon.

In order to make a successful transition between the SDGs and the MDGs, a few adjustments were needed. One of the strongest criticisms that the MDGs received

Table 1 Sustainable development goals

Sustainable development goals	
Goal 1	End poverty in all its forms everywhere
Goal 2	End hunger, achieve food security and improved nutrition and promote sustainable agriculture
Goal 3	Ensure healthy lives and promote well-being for all at all ages
Goal 4	Ensure inclusive and equitable quality education and promote lifelong learning opportunities for all
Goal 5	Achieve gender equality and empower all women and girls
Goal 6	Ensure availability and sustainable management of water and sanitation for all
Goal 7	Ensure access to affordable, reliable, sustainable and modern energy for all
Goal 8	Promote sustained, inclusive and sustainable economic growth, full and productive employment and decent work for all
Goal 9	Build resilient infrastructure, promote inclusive and sustainable industrialization and foster innovation
Goal 10	Reduce inequality within and among countries
Goal 11	Make cities and human settlements inclusive, safe, resilient and sustainable
Goal 12	Ensure sustainable consumption and production patterns
Goal 13	Take urgent action to combat climate change and its impacts
Goal 14	Conserve and sustainably use the oceans, seas and marine resources for sustainable development
Goal 15	Protect, restore and promote sustainable use of terrestrial ecosystems, sustainably manage forests, combat deseltification, and halt and reverse land degradation and halt biodiversity loss
Goal 16	Promote peaceful and inclusive societies for sustainable development, provide access to justice for all and build effective, accountable and inclusive institutions at all levels
Goal 17	Strengthen the means of implementation and revitalize the global partnership for sustainable development

Source United Nations (2015)

was their heightened focus on developing countries, which consequently brought extremely dependence on funding from rich countries (Kumar et al. 2016). Thus, the idea behind the SDGs was to become global, presenting a set of goals that are broader and can reach all countries (Ferreira 2013, Osborn et al. 2015). They have revolved around reaching prosperous lives, security, sustainable food and water, universal use of clean energy, healthy and productive ecosystems and global governance (Griggs et al. 2013). According to Osborn et al. (2015, p. 2) "all of the goals and targets contain important messages and challenges for developed and developing countries alike". Despite this, countries have the openness to select specific targets for the national scenario and determine their priorities and level of ambition for each objective (Allen et al. 2016).

Waage et al. (2015) highlight the scope of the Goals within different sectors, stating that they can be divided into areas of social welfare, infrastructure, environment and global governance, with intersectoral relationships and interdependence among all. However, the Goals are integrated, interdependent and achieved together, for instance, efforts to meet the education Goal (4) would contribute to poverty reduction (1) and economic growth (8) (Nilsson et al. 2016).

Also considering their interdependence, one goal achieved in an unsustainable way may prevent the achievement of another, e.g. using strategies to end poverty that promote unsustainable production and consumption, leads to failure in the achievement of goals that address sustainable production and consumption (Stafford-Smith et al. 2016). Stafford-Smith et al. (2016) further state that integration must be done by involving the interaction between sectors, countries, actors and multi-stakeholder partnerships, through what is envisaged in objective 17.

Gupta and Vegelin (2016) consider the SDGs as inclusive, bringing social and environmental issues into all the objectives, in an integrated way. Le Blanc (2015) also highlights the integration of the Goals, stating that they function in the form of a network, with targets that refer to multiple Goals at the same time. Thus, the SDGs becomes a set of integrated objectives, with targets to be met in the period from 2015 to 2030, aiming at ensuring fair and environmentally, economically and socially sustainable states.

This paper aims to discuss the role Education for Sustainable Development (ESD) has in assisting the Goals set by the 2030 Agenda for Sustainable Development. The next session will follow a literature review on ESD, exploring its emergence and definition. Later, session three will present a discussion on all the SDGs and how ESD can approach them. In order to explore the state of the art and scientific productions for ESD under the SDGs spectrum and the connection between them, three main databases were selected: Scopus, Web of Science and Science Direct. The main criteria for selecting the documents were: a time frame of the last eleven years of registration, in records from 2005, when the UN Decade of Education for Sustainable Development was launched, to 2016; to reflect the main ideas regarding ESD, environmental justice, sustainable development and the SDGs. In order to complete the collection of articles selected through the literature review and endorse the theoretical support of the research, an additional bibliographic research was carried out, using also books, UN reports, academic papers and journals cited in the selected publications of the databases.

2 Education for Sustainable Development

The Brundtland Commission Report, "Our Common Future" emphasized the importance of cooperation among the various stakeholders at the regional, national and global level as a precondition towards a sustainable future. The Rio Declaration on Environment and Development presented the goal of "establishing innovative equitable global partnerships through the creation of new levels of cooperation among States, key sectors of societies, and people" (United Nations 1992). Agenda 21, released at the Rio Summit, addressed the potential of the scientific and the technological community to make an effective contribution to the decision-making (Karatzoglou 2012).

The United Nations' Rio +20 Conference also brought the subject to light. In its final report "The Future We Want", the need to bring skills for advancing sustainable development for students is emphasized (United Nations 2012).

Sustainable Development originates from the three dimensions of the term, namely the environmental, economic, and sociocultural, what increases the need for extensive collaboration among diverse partners to effectively pursue its goals. The holistic approach, necessary to attain this transition to a sustainable society and simultaneously serve all three pillars, calls for a development marked by an increased complexity. Education Institutions have been generally considered significant contributors to the promotion of sustainability (Karatzoglou 2012).

Sustainable development was defined by the Brundtland Commission's report as "[…] development that meets the needs of the present without compromising the ability of future generations to meet their own needs" (Brundtland Commission Report 1987). In addition, sustainable development must be a consequence of actions related to economic, social and environmental development, not only in isolation, but integrated (Dvořáková and Zborková 2014).

Since Environmental Education (EE) was first introduced into the school curricula, it has struggled to establish its own identity. Environmental education started to be considered a concept in its own right up to the 1970s, focusing on the importance to develop environmental awareness to the society, before, it was dispersed in a diversity of disciplines that use the environment as a vehicle for teaching (Huiying 2002; McKeown and Hopkins 2003). The 1980s were important years for environmental education, in that it was a decade in which public environmental concern continued to heighten, giving environmental education a stronger impetus in schools. This decade was also significant since environmental education's holistic philosophy began to take root (Tilbury 1995).

Environmental education has the role of considering the environment in its totality (both social and natural aspects), aiming to build a continuous process through all phases of education, applying an interdisciplinary approach, looking at the local, the regional and the global facets, showing the need of cooperation (Kimaryo 2011). EE also brings the historical vision of the environmental issues to find solutions to prevent and to remedy current and future problems (Najam et al. 2007). The EE approach goes into the roots of symptoms and real reasons of environmental problems, highlighting its complexity and new paths to solve the issues. The methodology is diverse, using different education spaces to communicate and to educate about the environmental (Dias 2010).

Education for sustainable development was established as the central goal of environmental education in the 1990s: Sustainable living must be the new pattern for all levels: individuals, communities, nations and the world (United Nations 1992). To adopt the new pattern will require a significant change in attitudes and practices of many people. We will need to ensure that education programmes reflect the importance of an ethic for living sustainably (Tilbury 1995).

According to Unesco (2016) "Education for Sustainable Development allows every human being to acquire the knowledge, skills, attitudes and values necessary to shape a sustainable future". Education for Sustainable Development means including key sustainable development issues into teaching and learning; for example, climate change, disaster risk reduction, biodiversity, poverty reduction, and sustainable consumption (Wiek et al. 2011). It also requires participatory teaching and learning methods that motivate and empower learners to not only acquire knowledge, but also change their behavior and take action for sustainable development (McIntosh 2013; Luppi 2011; Teixeira 2013; Constantinescu 2014).

Education for Sustainable Development consequently promotes competencies like critical thinking, imagining future scenarios and making decisions in a collaborative way (Kibbe et al. 2014; Frantz and Mayer 2014; Zsóka et al. 2013; Carleton-Hug and Hug 2010). According to Unesco (2005, p. 18), ESD features are:

- Interdisciplinary and holistic: learning for sustainable development embedded in the whole curriculum, not as a separate subject;
- Values-driven: it is critical that the assumed norms—the shared values and principles underpinning sustainable development—are made explicit so that can be examined, debated, tested and applied;
- Critical thinking and problem solving: leading to confidence in addressing the dilemmas and challenges of sustainable development;
- Multi-method: word, art, drama, debate, experience, … different pedagogies which model the processes. Teaching that is geared simply to passing on knowledge should be recast into an approach in which teachers and learners work together to acquire knowledge and play a role in shaping the environment of their educational institutions;
- Participatory decision-making: learners participate in decisions on how they are to learn;
- Applicability: the learning experiences offered are integrated in day to day personal and professional life.
- Locally relevant: addressing local as well as global issues, and using the language(s) which learners most commonly use.

This coexistence of EE and ESD has created a concern among international and national communities regarding overlap and duplication of goals and programs in EE and ESD. Pavlova says that,

> The simultaneous existence and development of EE and ESD has resulted in some confusion in policy formulation and implementation. Lack of clarity has sometimes led to inefficiencies in achieving goals and development of initiatives. Some countries call for distinctions, others for convergence between the EE and ESD (Pavlova 2012, p. 333).

An analysis made by Unesco in 2009 concluded that the relationships between EE and ESD depends on "the historic role EE has played in a country (prominent or marginal) and the way EE itself is interpreted (broad or narrow)" (Unesco 2009a, b, p. 28). It identified that EE-ESD relationships can be considered as: equals; EE as a part of ESD; and ESD and EE as distinct, but considers them both important and that they end up overlapping (Unesco 2009a, b; Pavlova 2012). For the purpose of this paper, the authors will consider Environmental Education as part of Education for Sustainable Development.

With Agenda 21 emphasizing that education will play a major role in achieving sustainable development, ESD has received increasingly political attention. Its importance has been globally recognized with the establishment of the United Nations Decade for Education for Sustainable Development (DESD). Unesco (2013, p. 3), as the Decade's lead agency, defines ESD as "a process of learning how to make decisions that consider the long-term future of the economy, ecology and equity of all communities." For Haigh (2005, p. 31), the DESD "offers academies the best chance to date for making the deep and radical changes that will be necessary if the world's higher education institutions are to enact their responsibilities for creating a better and self-sustainable world."

Universities have often been considered major contributors of regional sustainability initiatives (Karatzoglou 2012; Dlouhá et al. 2011). The composite nature of Sustainable Development has appointed Universities to critical partners to all relevant efforts, always in firm collaboration with other local actors and promoting digital access to information (Dariah-EU 2016).

The study of moral reasoning in relation to sustainable development and environment is an emerging field within EE and ESD. Most environmental issues can be considered to be social dilemmas involving sophisticated moral conceptions, centered on notions of rights, freedoms, justice, equality and respect and this affects the academic, professional and moral formation of populations that are inserted in the paradigm of environmental protection (Unesco 2012; Kopnina 2014). Highlighting the ways in which environment benefits humans, recent articles call for humanizing education (Kopnina 2014; Wals 2012; Bowers 2002).

The integration of ESD in Higher Education implies shifts in current pedagogical strategies moving from: transmissive to discovery learning; teacher-centred to student-centred approaches; and theoretical to practice oriented learning (Lim et al. 2015). ESD is action orientated, based on critical reflection, including critical thinking, participatory approaches, partnerships and systemic thinking as its core. Existing teaching and learning approaches on ESD are based on educators as role models, experiential learning and holistic thinking (Unesco 2010). Besides transformative sustainability learning has been widely discussed in the literature as a suitable pedagogy in ESD (Cebrian 2014).

Action orientated approaches provide staff with the time to critically reflect on the practice of ESD and empower them to make changes in their teaching practice by acquiring new understandings and views on ESD (Wals 2015). Placing action research within the critical theory and emancipatory paradigm can contribute to transform participants' mental models on ESD. Furthermore, to engage academics in ESD it is necessary to develop a clearer rationale for the adoption of ESD pedagogy in Higher Education and conduct further empirical research on the views, practices and visions of those academics not involved in ESD (Cebrian 2014).

3 The Role of Education for Sustainable Development in the Sustainable Development Goals

The 2030 Agenda has presented ambitious goals for our future society. Carrying a strong sense of human development, human rights and seeking equity in all levels, the SDGs are an attempt to guide a more inclusive and just society. That is, sustainable development should be a balance of economic progress and the protection of the environment while also being mindful of social needs (Emas 2015; Dempsey et al. 2011). Umoh (2010, p.81) addresses the importance of education to this process, stating that "In a non quantifiable way, EE improves people's overall quality of life by opening opportunities for participation in development and processes, including the social, economic, political, and cultural spheres of life."

There are a number of authors that agree on the importance of education to change behaviors and promote critical thinking towards a more sustainable future (Kopnina 2015; Figueiró and Raufflet 2015; Kibbe et al. 2014; Frantz and Mayer 2014; Zsóka et al. 2013). Thus, Education for Sustainable Development is key to develop those skills in a transdisciplinary format that can help students realize that the pathways towards sustainability involves economical, political and social aspects. The challenge for educational institutions is precisely to reorganize their approach in a manner to align with the sustainability agenda (Jones et al. 2008). What will be discussed in this session are the SDGs and how is it feasible to approach them through education, particularly through ESD.

Goal 1 of the SDGs correspond to "No poverty", and it requires severe changes in national policies to guarantee fair distribution of resources and equitable development. Nonetheless, education also has an important role in the efforts to reduce poverty, or at least, in its effort to guide vulnerable groups of people on how to make more sustainable decisions. Climate change and its consequences currently has a stronger impact on low income families and ethnic groups, and those individuals are frequently lacking access to information, which makes them unable to seek relief assistance and understand their opportunities (Maantay and Maroko 2009).

Educational institutions must be prepared to deal with the challenges society is facing with climate change, so that it can shape their students to make conscious decisions. According to Power and Maclean (2011, p. 7) "Learning how to live and work in ways that are sustainable includes, but necessarily goes beyond, formal programs for education for sustainable development (ESD): the principles of sustainable development need to be installed in all levels and to cover all types of education."

In addition to the role of ESD to educate and inform the marginalized groups of society regarding climate change, there is also the role of capacity building. The knowledge and skills gained through education leads to more capacitated individuals that will have better chances of being employed or self-employed, consequently adding to the economic growth of their nation (Umoh 2010).

In this regard, the Food and Agriculture Organization of the United Nations (FAO), has published a report that addresses the importance of redefining education for rural people, in order to promote sustainable rural development, poverty reduction and food security. "Enhanced human capital in the rural space can be trained for increased onfarm productivity and for off-farm employment opportunities as well as learning that leads to improved social well-being, social capital formation and satisfactory livelihoods" (FAO 2009, p. 109). ESD is only one of the aspects to reduce poverty, however, its impact has long term effects that may not just benefit low income individuals, but society as a whole, promoting sustainable growth, which is also aimed by Goal 8, regarding Decent Work and Economic Growth.

The second Goal is highly connected with the first, and it concerns "Zero Hunger". In addition to eradicating hunger, the targets set by Goal 2 also aim to achieve food security and sustainable agriculture production. Within this context, target number three seeks to:

> […] double the agricultural productivity and incomes of small-scale food producers, in particular women, indigenous peoples, family farmers, pastoralists and fishers, including through secure and equal access to land, other productive resources and inputs, knowledge, financial services, markets and opportunities for value addition and non-farm employment (United Nations 2015).

By increasing food productivity and offering equal access to resources to small-scale producers, the SDGs are not only helping those low income groups to enhance their well-being, but also contributing to food security. Hence the connection between Goal 1 and 2 and the role ESD has on both issues. One of the characteristics of ESD is providing participatory learning (Ghilardi-Lopes et al. 2013; Barth et al. 2013; Soykan and Atasoy 2012) and practical activities (Kopnina 2015; Koscielniak 2014; Lozano et al. 2015) as a way to engage students to take what they have learn and put it into practice to understand better how their actions have consequences.

On the matter of food security, the promotion of agroecological practices in educational institutions are strongly supported. Through experimentation and observation of agroecological practices that fit the local environment enables the learners to acquire adaptive management skills which are key in dealing with climate change and its impact in agriculture and food security (Gregory et al. 2015). As presented by Jones (2013), an edible garden in the University of Gloucestershire in the UK counted not only with students, but local residents, staff and local government, which suggest that the impact of these types of practical activities can go beyond the campus if supported by the institution.

By motivating practical activities such as the edible garden, the time spent in nature and the good nutrition habits might be important drivers of good health and well-being, which is Goal 3. Besides, ESD can assist particularly in the achievement of target 3.9, that aims to "substantially reduce the number of deaths and illnesses from hazardous chemicals and air, water and soil pollution and contamination" (United Nations 2015) by creating awareness on the subject and teaching how to prevent those environmental issues.

Still on the matter of well-being, Goal 6, Clean Water and Sanitation can also be promoted through critical thinking, problem-solving skills, and by exercising environmental sensitivity towards the learner's community. Through outreach programs, the students can reach out to their own community, passing on knowledge and contributing to improve the quality of life of the region. According to EETAP (1997), "Students empowered to solve problems in their own neighborhoods will mobilize the communities to negotiate the issues and help them in communicating their opinions to the policy makers and community leaders."

Goal 4, Quality Education, wants to ensure that boys and girls, in every part of the world, can achieve literacy and improve their life by providing the skills necessary for their sustainable living. Target 4.7 especially focuses on ESD:

> By 2030, ensure that all learners acquire the knowledge and skills needed to promote sustainable development, including, among others, through education for sustainable development and sustainable lifestyles, human rights, gender equality, promotion of a culture of peace and non-violence, global citizenship and appreciation of cultural diversity and of culture's contribution to sustainable development (United Nations 2015).

This emphasizes the importance of education for sustainable development to the spread of a series of competences and knowledge that are key to pursue sustainable development. Educational institutions are a rich environment with the right tools to provide an education experience that is more meaningful to students (Wals et al. 1990). For Zenelaj (2013, p. 227) "Only well mannered people, have the power of speech and opinion to make everyone accountable, and to orient them on the path of sustainable development." At the same time, target 3.7, along with other targets from Goal 4 seeks to promote gender equality, which is also one of the SDGs, Goal 5.

Gender equality is one major characteristic of a democratic education process, along with social inclusion and diversity promotion, and education institutions have high potentials to promote those through offering equal opportunities to capacity building and access to knowledge (Gurin et al. 2002; Gómez et al. 2015). Besides, a research held by Unesco (2009a, b, p. 5) stated that "educating girls and women is one of the best ways of strengthening community adaptation to climate change." The premise of social inclusion also embraces Goal 8, "Decent Work and Economic Growth", Goal 9, "Industry, Innovation and Infrastructure", Goal 10, "Reduced Inequalities", Goal 11, "Sustainable cities and Communities", and Goal 16, "Peace, Justice and Strong Institutions."

On this matter, it is worth mentioning Target 10.2 of Goal 10, that aims to "empower and promote the social, economic and political inclusion of all, irrespective of age, sex, disability, race, ethnicity, origin, religion or economic or other status" (United Nations, 2015). A research by Vons et al. (2014) presented, through a case study in the south region of Brazil, a sensory garden as a way to promote social inclusion of deaf students within an environmental education program. The study concluded that the activity was successful, as it enabled the deaf students to exercise other senses other than their vision, which they are already familiarized with, but also contributed to the construction of scientific knowledge while respecting the students background and their ideas (Vons et al. 2014).

Goal 7 focuses on Affordable and Clean Energy, and has a strong connection with the development of new technologies. Jennings (1997, p. 2), affirms that education plays a major role in the renewable energy industry, because:

1. It promotes greater public awareness of the technology and confidence in using it.
2. The availability of technical advice and support services is essential for industry.
3. Education should be combined with a demonstration program to build confidence and awareness of the technology.
4. Education can provide a pool of trained researchers to continue the development of the technology.
5. Trained technical support staff are essential for installing, repairing and maintaining renewable energy systems.
6. People need to be aware of the technology and have the confidence to use it where appropriate.

Target 7.4 aims, "By 2030, enhance international cooperation to facilitate access to clean energy research and technology, including renewable energy, energy efficiency and advanced and cleaner fossil-fuel technology, and promote investment in energy infrastructure and clean energy technology." ESD has the potential to offer a much broader training in essential issues and to understand the development of new technologies that must be addressed by future professionals of the energy sector, which also results in new employment opportunities for those capacitated individuals (Jennings 2009; Jennings and Lund 2011).

As previously mentioned, one of the characteristics of ESD is behavior change. The education process must assume its role to prepare conscious and responsible citizens that will be capable of embracing what they have learned into their future actions. Thus, when it comes to Goal 12, "Responsible Consumption and Promotion", those Education Institutions who choose to incorporate Education for Sustainable Development into their curriculum will be exercising the learner's skills to "do more with less", which is a valuable tool to future challenges that climate change adaptation will require.

According to Tapia-Fonllem et al. (2013, p. 720) "a person that practices sustainable behavior not only engages in one kind of actions but tends to act in an integrated pro-environmentally manner." Therefore, simulation activities, which are often used in ESD, are a great example to reproduce first-hand experiences to facilitate the students perception on the aftermath of their actions. In addition, it can motivate independent, proactive, logical and creative behavior (Mingazova 2014; Steiner and Posch 2006). According to the Quality Assurance Agency for Higher Education, QAA (2014), simulation activities are also an important tool in developing appropriate professional behaviour given the possibility to use a safe environment to engage with real issues and different contexts.

The following SDGs, "Climate Action" (Goal 13), "Life Below water" (Goal 14), and "Life on Land" (Goal 15) are strongly linked given the impact of climate change on food production and biodiversity loss. Nonetheless, it is worth mentioning that Goal 13 is directly or indirectly related to the accomplishment of almost all the other goals (Ki-moon 2016). In this context, ESD is not only substantially relevant to awareness building. It has also the potential to introduce regional, national and international issues through a transdisciplinary form, enabling learners to identify different contexts and different point of views to problem-solving.

According to Unesco (2007, p. 6), "ESD prepares people to cope with and find solutions to problems that threaten the sustainability of the planet." At the same time, Steiner and Posch (2006) and Scholz et al. (2006) agree on the use of transdisciplinary practical activities as an effective mutual learning format to exercise problem-solving skills. Addressing climate change cannot be delayed any further and a learning environment is the most sensible place to engage debates, share new ideas and work together for a common goal. In addition, Unesco (2009a, b) believes that the involvement of the educational institution goes beyond the classroom, whereas the access to information on the subject helps to gain public trust on climate change related policies.

The last Goal, "Partnerships for the Goals", clarifies that there are a number of actors that must be involved in order to achieve all of the SDGs, including civil society, governments and the private sector. In this context of partnerships, educational institutions have the opportunity to join forces with the United Nations and use its platform to promote sustainable development. According to Unesco (2016), "Education alone cannot achieve a more sustainable future; however, without education and learning for sustainable development, we will not be able to reach that goal".

The seventeen goals here presented can all be inserted, in one way or another, to an educational context. The role of Education for Sustainable Development and the role of the Sustainable Development Goals has many parallels and that must be exploited to the benefit of society. As highlighted by the last Goal, partnerships are key to this endeavor, thus, they should be promoted. At the same time, Goal 4 explicitly puts ESD as a component to promote sustainable development. What remains is the will of education institution leaders to seize this opportunity to adjust their curricula and prepare the teaching and student body to embrace this challenges.

4 Final Remarks

Agenda 2030 emerges as the journey forward after the Millennium Development Goals final deadline. Learning from the gains and halts of the MDGs, the Sustainable Development Goals made an effort, during the three years of its negotiation, to include diversified views and a greater participation of civil society. From the beginning it became undeniable that partnerships were necessary, not only during the development process of the Goals but primarily for their implementation.

Throughout this paper, Education for Sustainable Development is presented as a possible partner and articulator of the SDGs.

The UN Decade of Education for Sustainable Development was a major step towards a structural change in education systems. The concept, which arose from Environmental Education, was built upon the need of offering a broader definition on the meaning of educating about the environment, also focusing on economic and social issues in order to approach the three pillars of sustainable development. Thus, following the ESD approach means addressing the same issues exposed in the SDGs.

As a platform for debates and knowledge, that shares the same dilemmas such as seeking equity and addressing local and global issues, ESD has a great potential to be a driver of the SDGs albeit following its own methodology. Session three exposed some of the characteristics of ESD such as practical activities, interdisciplinary and transdisciplinary approach, problem-solving skills and environmental sensitivity towards the learner's community. These are just a few of the possible outreach potentials that ESD has to address the goals and targets set by Agenda 2030 through an inclusive and engaging manner.

The quest for sustainable development does not concern only one group of people nor does it concern only one agenda. Preparing people for future challenges like climate change and resource depletion requires a society with the knowledge and skills to act on the issues surrounding their own space and for that, education is the only way. Nonetheless, the pathway towards the accomplishment of the SDGs is still a lengthy one and it depends on many variables. Education for Sustainable Development is only one of the actors of this endeavor, but it is a very important one.

Acknowledgements This study was conducted by the Energy Efficiency and Sustainability Research Group (GREENS) from the University of Southern Santa Catarina in the context of the Projects: Links 2015—Linkages between energy, food and water consumption in the context of climate change mitigation strategies, and BRIDGE—Building Resilience in a Dynamic Global Economy: Complexity across scales in the Brazilian Food-Water-Energy Nexus, funded by the Newton Fund and FAPESC—Fundação de Amparo à Pesquisa e Inovação do Estado de Santa Catarina.

References

Allen, C., Metternicht, G., & Wiedmann, T. (2016). National pathways to the sustainable development goals (SDGs): a comparative review of scenario modelling tools. *Environmental Science and Policy, 66,* 199–207. doi:10.1016/j.envsci.2016.09.008.

Barth, M., Adomßent, M., Fischer, D., Richter, S., & Rieckmann, M. (2013). Learning to change universities from within: a service-learning perspective on promoting sustainable consumption in higher education. *Journal of Cleaner Production, 62,* 72–81. doi:10.1016/j.jclepro.2013.04.006.

Bowers, C. (2002). Toward an eco-justice pedagogy. *Environmental Education Research, 8*(1), 21–34. doi:10.1080/13504620120109628.

Brundtland Commission Report. (1987). Report of the world commission on environment and development: Our common future. http://www.un-documents.net/our-common-future.pdf.
Carleton-Hug, A., & Hug, J. W. (2010). Challenges and opportunities for evaluating environmental education programs. *Evaluation and Program Planning, 33,* 159–164. doi:10.1016/j.evalprogplan.2009.07.005.
Cebrian, G. (2014). *Academic Staff Engagement in Education for Sustainable Development.* doi:10.1016/j.jclepro.2014.12.010.
Constantinescu, C. (2014). Valuing interdependence of education, trade and the environment for the achievement of sustainable development. *Procedia—Social and Behavioral Sciences, 116,* 3340–3344. doi:10.1016/j.sbspro.2014.01.760.
Dariah-EU. (2016). CFP: Digital Editions - Representation, Interoperability, Text Analysis and Infrastructures. Digital Research Infrastructure for the Arts and Humanities. http://dariah.eu/news/articles.html.
Dempsey, N., Bramley, G., Power, S., & Brown, C. (2011). The social dimension of sustainable development: defining urban social sustainability. *Sustainable Development, 19,* 289–300. doi:10.1002/sd.417.
Dias, G. F. (2010). Educacao Ambiental, princípios e práticas, 9ª edicao, Ed Gaia, 2010.
Dlouhá, J., Adomssent, M., Barton, A., & Fadeeva, Z. (2011). Learning for sustainable development in regional networks. *Journal of Cleaner Production, 19,* 1144–1145. doi:10.1016/j.jclepro.2011.02.002.
Dvořáková, L., & Zborková, J. (2014). *Integration of Sustainable Development at Enterprise Level.* doi:10.1016/j.proeng.2014.03.043.
EETAP. (1997). Environmental education for empowerment. EETAP Resource Library, Advancing Education & Environmental Literacy, 13. http://resources.spaces3.com/df7372d0-7e77-4779-9004-4623587bfcf8.pdf.
Emas, R. (2015). The concept of sustainable development: definition and defining principles. Brief for GSDR 2015. https://sustainabledevelopment.un.org/content/documents/5839GSDR%202015_SD_concept_definiton_rev.pdf.
FAO. (2009). Education for rural people. The role of education, training and capacity development in poverty reduction and food security. http://www.fao.org/3/a-i0760e.pdf.
Ferreira, P. M. (2013). A Agenda Pós-2015 para o desenvolvimento:Da redução da pobreza ao desenvolvimento inclusivo? Policy Paper, 1/2013, 2–21. http://www.imvf.org/ficheiros/IMVFPolicyPaper_pos2015.pdf.
Figueiró, P., & Raufflet, E. (2015). Sustainability in higher education: a systematic review with focus on management education. *Journal of Cleaner Production, 106,* 22–33. doi:10.1016/j.jclepro.2015.04.118.
Frantz, C. M. P., & Mayer, F. S. (2014). The importance of connection to nature in assessing environmental education programs. *Studies in Educational Evaluation, 41,* 85–89. doi:10.1016/j.stueduc.2013.10.001.
Ghilardi-Lopes, N. P., Braga, J. C., Freitas, S.R., Silva, R. L. F., Ribeiro, T., Simões, E., Pinhata, E., & Silva, L. (2013). Environmental education through an online game about global environmental changes and their effects on coastal and marine ecosystems. In *Conference proceedings of SBGames* (pp. 468–474). http://dx.doi.org/10.1590/S1679-875920160932064sp2.
Gómez, F. U., Sáez-Navarrete, C., Lioi, S. R., & Marzuca, V. I. (2015). Adaptable model for assessing sustainability in higher education. *Journal of Cleaner Production, 107,* 475–485. doi:10.1016/j.jclepro.2014.07.047.
Gregory, M. M., Leslie, T. W., & Drinkwater, L. E. (2015). Agroecological and social characteristics of New York city community gardens: Contributions to urban food security, ecosystem services, and environmental education. *Urban Ecosystem, 19,* 763–794. doi:10.1007/s11252-015-0505-1.
Griggs, D., Stafford-Smith, M., Gaffney, O., Rockström, J., Öhman, M. C., Shyamsundar, P., et al. (2013). Policy: sustainable development goals for people and planet. *Nature, 495*(7441), 305–307.

Gupta, J., & Vegelin, C. (2016). Sustainable development goals and inclusive development. In *International environmental agreements: Politics, law and economics* (pp. 1–16).

Gurin, P., Dey, E., Hurtado, S., & Gurin, G. (2002). Diversity and higher education: Theory and impact on educational outcomes. *Harvard Educational Review, 72,* 330–367. doi:10.17763/haer.72.3.01151786u134n051.

Haigh, M. (2005). Greening the university curriculum: appraising an international movement. *Journal of Geography in Higher Education, 29,* 31–48. doi:10.1080/03098260500030355.

Haslegrave, M. (2014). Sexual and reproductive health and rights in the sustainable development goals and the post-2015 development agenda: Less than a year to go. *Reproductive health matters, 22*(44), 102–108.

Huiying, X. (2002). Interdisciplinary approaches to environmental education: Two case studies in China. http://www.cgie.org/blog/resources/papers-publications/interdisciplinary-approaches-environmental-education-case-studies-china/.

Jennings, P. J. (1997). Renewable energy education: an essential foundation for market development. In *Proceedings of Solar 97—Australian and New Zealand solar energy society*, Paper 33. http://researchrepository.murdoch.edu.au/25033/1/renewable_energy_education.pdf.

Jennings, P. (2009). New directions in renewable energy education. *Renewable Energy, 34,* 435–439. doi:10.1016/j.renene.2008.05.005.

Jennings, P. J., & Lund, C. P. (2011). Renewable energy education for sustainable development. *Renewable Energy, 22,* 113–118. doi:10.1016/S0960-1481(00)00028-8.

Jones, D. R. (2013). 'The Biophilic University': a de-familiarizing organizational metaphor for ecological sustainability? *Journal of Cleaner Production, 48,* 148–165. doi:10.1016/j.jclepro.2013.02.019.

Jones, P., Trier, C. J., & Richards, J. P. (2008). Embedding education for sustainable development in higher education: a case study examining common challenges and opportunities for undergraduate programmes. *International Journal of Educational Research, 47,* 341–350. doi:10.1016/j.ijer.2008.11.001.

Karatzoglou, B. (2012). An in-depth literature review of the evolving roles and contributions of universities to education for sustainable development. doi:10.1016/j.jclepro.2012.07.043.

Kibbe, A., Bogner, F. X., & Kaiser, F. G. (2014). Exploitative vs. appreciative use of nature—Two interpretations of utilization and their relevance for environmental education. *Studies in Educational Evaluation, 41,* 106–112. doi:10.1016/j.stueduc.2013.11.007.

Kimaryo, L. (2011). Integrating environmental education in primary school education in Tanzania teachers' perceptions and teaching practices. http://www.doria.fi/bitstream/handle/10024/67481/kimaryo_lydia.pdf.

Ki-moon, B. (2016) The people's agenda. In *Sustainable development goals: The people's agenda.* UNA-UK.

Kopnina, H. (2014). Education for sustainable development (ESD): Exploring anthropocentric–ecocentric values in children through vignette. doi:10.1016/j.stueduc.2013.12.004.

Kopnina, H. (2015). Neoliberalism, pluralism and environmental education: the call for radical re-orientation. *Environmental Development, 15,* 120–130. doi:10.1016/j.envdev.2015.03.005.

Koscielniak, C. (2014). A consideration of the changing focus on the sustainable development in higher education in Poland. *Journal of Cleaner Production, 62,* 114–119. doi:10.1016/j.jclepro.2013.06.006.

Kumar, S., Kumar, N., & Vivekadhish, S. (2016). Millennium development goals (MDGS) to sustainable development goals (SDGS): Addressing unfinished agenda and strengthening sustainable development and partnership. *Indian Journal of Community Medicine (Editorial Commentary), 41,* 1–4. doi:10.4103/0970-0218.170955.

Le Blanc, D. (2015). Towards integration at last? The sustainable development goals as a network of targets. *Sustainable Development, 23*(3), 176–187.

Lim, Y. S., Xia, B., Skitmore, M., Gray, J., & Brodge, A. (2015). Education for sustainability in construction management curricula. *International Journal of Construction Management, 15,* 321–331. doi:10.1080/15623599.2015.1066569.

Lozano, R., Ceulemans, K., Alonso-Almeida, M., Huisingh, D., Lozano, F., Waas, T., et al. (2015). A review of commitment and implementation of sustainable development in higher education: results from a worldwide survey. *Journal of Cleaner Production, 108,* 1–18. doi:10.1016/j.jclepro.2014.09.048.

Luppi, E. (2011). Training to education for sustainable development through e-learning. *Procedia Social and Behavioral Sciences, 15,* 3244–3251. doi:10.1016/j.sbspro.2011.04.279.

Maantay, J., & Maroko, A. (2009). Mapping urban risk: Flood hazards, race, & environmental justice in New York. *Applied Geography, 29,* 111–124. doi:10.1016/j.apgeog.2008.08.002.

McIntosh, M. (2013). Creating global citizens and responsible leadership. The Journal of Corporate Citizenship, Griffith University, Issue 49, Australia. Available at https://www.greenleaf-publishing.com/creating-global-citizens-and-responsible-leadership.

McKeown, R., & Hopkins, C. (2003). EE p ESD: Defusing the worry. *Environmental Education Research, 9,* 117–128. doi:10.1080/13504620303469.

Mingazova, N. M. (2014). Modification Of the active learning methods in environmental education in Russian universities. *Procedia—Social and Behavioral Sciences, 131,* 85–89. doi:10.1016/j.sbspro.2014.04.083.

Najam, A., Runnalls, D., Halle, M. (2007). Environment and globalization five propositions. In *International Institute for Sustainable Development.* http://www.unep.org/gc/gc24/docs/FivePropositions.pdf.

Nilsson, M., Griggs, D., & Visbeck, M. (2016). Policy: map the interactions between sustainable development goals. *Nature, 534,* 320–322.

Osborn, D., Cutter, A., & Ullah, F. (2015). Universal sustainable development goals. Understanding the transformational challenge for developed countries. London: Report of a study Stakeholder Forum.

Pavlova, M. (2012). Environmental education and/or education forsustainable development: what role for technology education? http://www98.griffith.edu.au/dspace/bitstream/handle/10072/46566/74769_1.pdf?sequence=1.

Post-2015 HLP. Secretary-General's High-Level Panel of eminent persons on the Post-2015 Development Agenda. About. http://www.post2015hlp.org/about/.

Power, C., Maclean, M. (2011). Lifelong learning for poverty alleviation and sustainable development. In *International symposium on lifelong learning*. https://www.ied.edu.hk/isll/Download%20Files/Panel2_Prof_Colin_Power.pdf.

QAA. (2014). Education for sustainable development: Guidance for UK higher education providers. http://www.qaa.ac.uk/en/Publications/Documents/Education-sustainable-development-Guidance-June-14.pdf.

Sachs, J. D. (2012). From millennium development goals to sustainable development goals. *The Lancet, 379*(9832), 2206–2211.

Scholz, W. R., Lang, D. J., Wiek, A., Walter, A. I., & Stauffacher, M. (2006). Transdisciplinary case studies as a means of sustainability learning Historical framework and theory. *International Journal of Sustainability in Higher Education, 7,* 226–251. doi:10.1108/14676370610677829.

Soykan, A., & Atasoy, E. (2012). Historical development of non-formal environmental education in Turkey. *Procedia—Social and Behavioral Sciences, 46,* 736–743. doi:10.1016/j.sbspro.2012.05.190.

Stafford-Smith, M., Griggs, D., Gaffney, O., Ullah, F., Reyers, B., Kanie, N., Stigson, B., Shrivastava, P., Leach, M., & O'Connell, D. (2016). Integration: the key to implementing the Sustainable Development Goals. Sustainability Science (pp. 1–9).

Steiner, G., & Posch, A. (2006). Higher education for sustainability by means of transdisciplinar case studies: an innovative approach for solving complex, real-world problems. *Journal of Cleaner Production, 14,* 877–890. doi:10.1016/j.jclepro.2005.11.054.

Tapia-Fonllem, C., Corral-Verdugo, V., Fraijo-Sing, B., & Durón-Ramos, M. F. (2013). Assessing sustainable behavior and its correlates: A measure of pro-ecological, frugal, altruistic and equitable actions. *Sustainability, 5,* 711–723. doi:10.3390/su5020711.

Teixeira, S. R. (2013). The environmental education as a path for a global sustainability. *Procedia—Social and Behavioral Science, 106,* 2769–2774. doi:10.1016/j.sbspro.2013.12.318.
Tilbury, D. (1995). Environmental education for sustainability: Defining the new focus of environmental education in the 1990s. doi:10.1080/1350462950010206.
Umoh, T. U. (2010). Environmental education and poverty reduction in Africa. *Journal of African Business, 11,* 70–86. doi:10.1080/15228911003608561.
Unesco. (2005). UN decade of education for sustainable development 2005–2014: International Implementation Scheme. http://unesdoc.unesco.org/images/0013/001399/139937e.pdf.
Unesco. (2007). The US decade of education for sustainable development, the first two years. http://unesdoc.unesco.org/images/0015/001540/154093e.pdf.
Unesco. (2009a). Climate change and environmental education. In Child-friendly schools manual. https://www.unicef.org/publications/index_49574.html.
Unesco. (2009b). Learning for a sustainable world: Review of contexts and structures for education for sustainable development. Paris: UNESCO Education Sector. http://unesdoc.unesco.org/images/0018/001849/184944e.pdf.
Unesco. (2010). Training guideline on incorporating education for sustainable development (ESD) into the curriculum. http://www.ibe.unesco.org/sites/default/files/ESD_training_guidelines_-3.pdf.
Unesco. (2012). Shaping the education of tomorrow: 2012 full-length report on the UN decade of education for sustainable development. UNESCO Education Center. http://unesdoc.unesco.org/images/0021/002166/216606e.pdf.
Unesco. (2013). Education for sustainable development (ESD) in the UK—Current status, best practice and opportunities for the future. http://www.unesco.org.uk/wp-content/uploads/2015/03/Brief-9-ESD-March-2013.pdf.
Unesco. (2016). UN decade of ESD. http://en.unesco.org/themes/education-sustainable-development/what-is-esd/un-decade-of-esd.
United Nations. (1992). Rio declaration on environment and development. http://www.un.org/documents/ga/conf151/aconf15126-1annex1.htm.
United Nations. (2012). United Nations General Assembly."The Future We Want" Resolution adopted by the General Assembly on 27 July 2012. http://www.un.org/ga/search/view_doc.asp?symbol=A/RES/66/288&Lang=E.
United Nations. (2015). United Nations General Assembly. Transforming Our World: The 2030 agenda for sustainable development. https://sustainabledevelopment.un.org/post2015/transformingourworld.
UNSDSN. (2016a). United Nations Sustainable Development Solutions Network. About the SDGs. http://unsdsn.org/what-we-do/sustainable-development-goals/about-the-sdgs/.
UNSDSN. (2016b). United Nations Sustainable Development Solutions Network. Vision and Organization. http://unsdsn.org/about-us/vision-and-organization/.
Vons, P. C. O., Scopel, J. M., & Scur, L. (2014). The Sensory Garden as an Environmental Education Activity for Social Inclusion in the Natural Science Museum of University of Caxias do Sul. *Scientia Cum Industria, 2,* 43–46. doi:10.18226/23185279.v2iss2p43.
Waage, J., Yap, C., Bell, S., Levy, C., Mace, G., Pegram, T., et al. (2015). Governing the UN sustainable development goals: Interactions, infrastructures, and institutions. *The Lancet Global Health, 3*(5), e251–e252. doi:10.1016/S2214-109X(15)70112-9.
Wals, A. E. J. (2012). Shaping the education of tomorrow: 2012 Full-length Report on the UN decade of education for sustainable development. UNESCO Education Center. http://unesdoc.unesco.org/images/0021/002166/216606e.pdf.
Wals, Arjen. (2015). Vers une éducation plus significative et plus critique et l'apprentissage pour la durabilité - aller au-delà du Décennie de l'éducation pour le développement durable. http://dx.doi.org/10.1080/15623599.2015.1066569.
Wals, A. E., Beringer, A., & Stapp, W. B. (1990). Education in action: A community problem-solving program for schools. *The Journal of Environmental Education, 21,* 13–19. doi:10.1080/00958964.1990.9941933.

Wiek, A., Withycombe, L., Redman, C.L. (2011). Key competencies in sustainability: A reference framework for academic program development. Sustainability Science, 6. doi:10.1007/s11625-011-0132-6.

Zenelaj, E. (2013). Education for sustainable development. European Journal of Sustainable Development, 2, 227–232. http://www.ecsdev.org/images/conference/siICSD2013/zenelaj%20227-232.pdf.

Zsóka, A., Szerényi, Z. M., Széchy, A., & Kocsis, T. (2013). Greening due to environmental education? Environmental knowledge, attitudes, consumer behavior and everyday proenvironmental activities of Hungarian high school and university students. *Journal of Cleaner Production, 48,* 126–138. doi:10.1016/j.jclepro.2012.11.030.

Explicit Economics: Addressing Conscious Consumption for Sustainability

Madhavi Venkatesan

Abstract Sustainability is typically discussed in a siloed fashion in the United States. Cradle-to-cradle production and regulation to curb greenhouse gas emissions are proffered as salves for evidenced degradation but little attention is directed to how a society can enable sustainability as a cultural norm. Further and related, the role of the individual economic agent as consumer, investor, and government participant is seemingly not acknowledged. To a large extent, the population majority delegates the powers conferred in the three roles to a minority largely through indifferent conveyance posited on trust, leaving outcomes impacting society as a whole dependent on incentives of a few, who may or may not be aligned with the public welfare. Therefore, given the evidence of marketed demand fostered by a consumerism based economy (Day and Aaker in J Mark 34(3):12–19, 1970), perhaps the most significant, powerful, and traction-inducing vehicle for instituting sustainability may be found in enabling conscious consumption at the individual level. Arguably, the conduit for conscious consumption would then be education not limited to defining sustainability but inclusive of the rationale for sustainability, the patience requisite for implementation, and the acceptance of sustainability as a societal norm of behavior. However, the building block for conscious consumption is found in understanding the basis of present consumption decisions, ultimately the values that shape the behaviors that lead to observable economic outcomes.

Keywords Sustainability · Conscious consumption · Economics · Values

M. Venkatesan (✉)
Department of Economics, Bridgewater State University, 131 Summer Street, Bridgewater, MA 02325, USA
e-mail: mvenkatesan@bridegw.edu

J.P. Davim (ed.), *Curricula for Sustainability in Higher Education*,
Management and Industrial Engineering, DOI 10.1007/978-3-319-56505-7_2

1 Introduction: Significance of Educating for Sustainability

In the United States, consumption expenditures contribute to over 65% of gross domestic product (GDP), which since the 1940s has been the international metric for economic progress (Venkatesan 2015). Given this linkage and the corresponding focus on GDP growth as a proxy for progress, consumption decisions can have a significant ripple effect throughout a single economy as well as the finite global resource base. Consider for example the use of milk cartons. Wax lined, printed paper milk cartons have been created for the transport and preservation of milk from the production to the consumption stage. However, the components of the carton were not developed with waste disposal in mind, rather increasing distribution and sales were the rationale for the carton. As a result, largely related to the focused basis of its creation, the milk carton serves a consumption purpose without consideration of the impact to the environment and potential future human and animal health due to its non-biodegradable or re-usable composition. This illustration on a broader consumption scale provides a simplified perspective to evaluate the underlying values captured in consumption decisions (Schweitzer 1981; Colander 2005). From this perspective, production for consumption may be expressed as a myopic activity, focused on near-term satiation of a need or want to the exclusion of the evaluation of the impact or ripple effect of the satiation.

The values embedded and communicated within demand and supply, determine the manner in which a need or want is attained. To the extent that there is no discussion of the values and behavioral factors assumed and reflected in demand and supply, arguably, implicit values, the values and the subsequent behaviors become endogenous to the economic system. Therefore, explicit awareness of present behavioral assumptions inclusive of the "unlimited wants" of consumers, profit maximization motivations of producers, and the understated resource depletion resulting from externalized costs, offer the potential to modify active and embedded behavior.

An understanding of economics specifically oriented toward enabling the development of rational economic agent behavior, can raise awareness of the significance of consumption behavior as the activity relates to sustainability, where the defining of sustainability is consistent with the economic perspective of resource utilization bounded by intergenerational equity, which by definition considers the needs of the present and the future relative to current resource capacity and regeneration ability. Awareness of resource limitations in conjunction with temporal allocation, in turn fosters the development and implementation of conscious and unconscious reinforcement of sustainability, which are the needed elements in driving a culture of sustainability.

2 Economics in Cultural Context

Economics evaluates human behavior relative to wants, needs and resource allocation within a natural environment. By definition, the parameters of the discipline include other life forms and physical resources needed to maintain both life and environmental regeneration. To the extent that a human culture incorporates non-human elements in decision-making, the economic system includes an understanding of the holistic inter-dependence of living and non-living elements of the planet.

Culture is a significant contributor to what is perceived as valuable and is the determining parameter in the designations that ultimately yield to resource allocation within a society (O'Hara 1995). Given that culture is a learned behavior, culture can either promote or diminish any given society's understanding of the interconnectedness of human and planetary life, thereby determining the extent of the anthropocentric, or human-centered, perspective. The United Nations Educational, Scientific and Cultural Organization, UNESCO, defined culture as a significant component to attaining global sustainability:

> Culture shapes the way we see the world. It therefore has the capacity to bring about the change of attitudes needed to ensure peace and sustainable development which, we know, form the only possible way forward for life on planet Earth. Today, that goal is still a long way off. A global crisis faces humanity at the dawn of the 21st century, marked by increasing poverty in our asymmetrical world, environmental degradation and shortsightedness in policy-making. Culture is a crucial key to solving this crisis (UNESCO 2000).

The inputs and outputs of economic systems are dependent on the value structures of a society and to the extent that economics explains observable phenomenon and proposes optimal outcomes, the discipline can be both responsible for the maintenance of an economic framework and also the catalyst for a change. Economic outcomes in essence mimic the values of the participants in an economic system.

Evaluating the historical cultural progression of human society can promote a stronger understanding of the economic relationship with resource allocation, both intra- and inter-society, and most importantly provide insights with respect to how perceptions of the world are shaped through cultural frameworks at a given point in time. The pace at which cultural attributes evolve may also provide a deeper understanding of why institutional and social frameworks may be inconsistent with the manifestation of contemporary challenges. Viewing economic thought or philosophy over time reveals the dynamic and cultural elements of society, as well as the basis of economic thought that remains in the principles literature in the present period.

3 Reconciling Economic Theory and Historical Context

The cultural attribution of value is a significant and arguably primary differentiator with respect to the variation in the perspective between societies of the quality of life for both human and non-human elements. Examples of surviving written works that provide a foundation or insight with respect to economic activities include Plato's *Republic* and Aristotle's *Politics*. The similarities in economic circumstances as described by the authors are consistent with the phenomenon observable today; however, the evaluation of human behavior as it applied to accumulation of wealth, stratification of society, and the role and impact of gratification were framed within an evaluation and discussion of moral philosophy and ethics, positioning Western economics up to the eighteenth century within the discipline of moral philosophy and politics. The evolution of the discipline continued through the modern era until the discipline formerly separated from moral and political philosophy through iteration as political economy to its present standalone context as economics. The observable mechanics of economic systems were the basis of discussion in conjunction with the human values, whether assumed as innate or culturally inspired. A connection between the qualitative and quantitative aspects of economic outcomes was articulated and addressed as an evolving and dynamic process. From this perspective economics discussions offered both a *normative* and a *positive* perspective, where the former provided opinions and values related to optimization and the latter described observable activity. At the present time, economics in practice has shed the normative element of the discipline opting for a positive attribution as a means to enhance its standing as a science. In essence the focus on optimization has been to the exclusion of explicit evaluation of prevailing values. Given the significance of embedded values in conscious decision making, the lack of articulation of values may contribute to the implicit value of outcome based decision-making that only considers the optimization of the outcome rather than the impact of the outcome to others and future consumption.

The foundation for current economic thought can be found in the writings of Smith (1791), Bentham (1879), Ricardo (1911) and Marx (1968) along with many others. However, though all of these authors provided insights related to the human behavior contemporary to their time, the context of their writings has often been neglected in lieu of an adoption of an absolute meaning of their opinions. In essence, allowing the commentaries of these authors to embody a universal significance independent of time has arguably enabled the transfer of the theoretical modeling of a society specific from one period to another, independent of any assessment of the temporal evolution of behavior and underlying values.

To a large extent, the economic principles in practice have maintained the theories espoused by the writers and contributors to economic thought contemporary to the Classical period. Mill's (2016) *Principles of Political Economy* provided a summary of the contributions to economic thought by Adam Smith, David Ricardo and other significant thought leaders of the nineteenth century and became a standard text used in the study of economics into the early twentieth century.

However, of note is that the authors including Mill were relaying behaviors perceived in a society contemporary to their life and questioning aspects of the observed progress of the time including poverty, the role of money, and the potential impact of population growth. Their thoughts were debated discussions and their frameworks were not adopted as immutable facts. Additionally, the issues discussed were similar to those of predecessor Western societies and as evidenced in the moral philosophical discourses of Plato and Aristotle, nearly two millennia earlier. The evaluation of the human condition within a given social and economic framework prompted economic commentators to be both positive evaluators from the perspective that positive signifies reporting on observable and factual phenomenon and normative participants, where normative requires an expression of value judgment.

In contrast with the foundations of the discipline, the present instruction of economics has eliminated the normative aspects of assessment, reducing economics to mathematical relationships that are addressed in absolute terms rather than in alignment with cultural attributions coincident with their development. Further the seeming lack of attention to values and behavior incorporated within economic assessment has distanced the tangibility of economics, limiting understanding of the explanatory potential of economics and the application of economics as both a cause and a remedy of unsustainable practices. As a result, at present there is a need to promote and foster an understanding of the role of values in economic outcomes and the sustainability of observed outcomes.

4 Perception of Resource Value, Market Outcomes and Price

Economics is the social science discipline that evaluates the relationship between human wants and the resources available to satisfy them. In identifying and explaining the relationship between wants and resources, economists use broad generalizations related to human behavior, arguably the most significant of which relates to wants (Fagg 1981).

Wants are based on the premise that individual economic agents, individuals interacting within the general economy, will always seek to have more of desirable goods and services. Desirable goods include both normal goods, which are goods that an individual will continue to purchase as their income increases and luxury goods, which are goods that are not needed but are wanted to support an external display or perception of status or wealth. Not all goods are desirable, for example, inferior goods represent a classification of goods and services that will be reduced or eliminated by consumers as their incomes increase.

The behavior of wanting more, sometimes referenced as unlimited wants, is a social value, consistent with consumerism, which is defined as the focused act of consuming goods and services to improve utility, the economic concept that defines the benefit of consumption. Insatiability is not representative of an intrinsic human characteristic but rather a learned behavior. This is an important point. If a behavior is learned it can be unlearned and a new behavior can emerge, which in turn can produce a different economic outcome.

5 Market Distortions, Externalities and Failure of Market Equilibrium

Market outcomes, price and quantity, are highly dependent on the information that consumers and suppliers have available. Informational asymmetry, where one party has more understanding or knowledge related to a good than another party, can create price and quantity outcomes that may not effectively consider scarcity. This results in market inefficiency, a situation where resource use is not efficiently allocated by the market. This is a significant issue and one that consumers are only beginning to understand. For example, abundance is a relative term but it is not inconsistent with scarcity; all resources are scarce. The perception of abundance without the recognition of inherent scarcity of resources can hasten resource depletion.

Resources are broadly defined as including all the inputs in the production of final goods and services that are ultimately tied to the satisfaction of a want. From this perspective, resources could include teak wood trees in the making of furniture; water in the production of soda; and cattle in the production of food. Typically, resources are classified into one of three groupings, which include: natural resources, human resources, and capital resources. Trees, water, and cattle are all natural resources. Human labor or entrepreneurship define human resources and capital resources consist of man-made objects that can be used to produce goods and services, such as factories and equipment. Regardless of the type of resource, all resources are finite and so by definition can be qualified as scarce (Venkatesan 2016; Czech 2000).

Scarcity in economics essentially captures the relationship between wants and the access and availability of resources. For example, one could want a mango, see it hanging high on a tree but not have a ladder to reach it. The good in question is available but it is not accessible. Alternatively, one could stumble on a farmer's market selling mangos only to find that all the mangoes on display have been purchased. In this case the mangos are accessible but they are not available. Both of these examples highlight the temporal or time sensitivity of scarcity. In the first example, one could borrow or purchase a ladder but this will take time and in the second scenario, one can drive or walk to another market, but again, additional time will be required to satisfy the want.

Looking at time in a slightly different manner, a community could require lumber for the construction of new municipal buildings. The lumber required will result in the deforestation of one hundred acres. In satisfying the want for lumber today, the community limits access and availability of lumber from the one hundred acres over the time period required for the forest to regenerate, creating time based scarcity.

In a market system, access and availability establishes a perceived scarcity embedded within the supply of a good. Ultimately, the supplier's willingness and ability to sell a specified amount of a good at a prevailing price is assumed to capture the costs of production of the good, implicitly including the scarcity of inputs. As a result it is expected that the higher the degree of perceived scarcity of a resource, the higher its price and in the case of an input, the resulting price of the final good.

The production of goods by producers is based on a competitive framework. Additionally, the producer seeks to minimize costs and maximize revenue, to achieve maximum profitability. As a result of the focus on profitability, there is significant incentive for producers to externalize costs of production as a means of cost minimization. Externalizing costs can include pollution discharge, exploitation of regulatory differences between countries, overuse of natural resources, and limited waste disposal and reduction efficiencies. Though in the immediate period this may be beneficial to profitability, it may promote both short-lived unsustainable returns and longer-term environmental and social costs.

Consumers may not be aware of the implicit tradeoffs being made as a result of the production of a good. This informational asymmetry can be attributable to many reasons, including a belief that regulatory agencies guarantee safety, to just simply a lack of diligence when assessing goods. For consumers, reliance on market efficiency without an understanding of the embedded incentives of producers can promote negative externalities. In effect the pursuit of satisfying unlimited wants may include effectively delegating environmental and social stewardship to producers whose incentives may not include the evaluation of these parameters. The end result is most readily seen in natural resources, where under pricing (Boran 2006) due to lack of inclusion of scarcity can lead to extinction or elimination of a resource's availability.

In a market driven economy, such as in the United States, the market is credited with efficiently determining the price of an item by implicitly incorporating the costs associated with production. When consumers or producers face low prices for consumption and input purchases, respectively, and the underlying belief is that the price being paid is fully reflective of the cost of the item being purchased, there is less of an incentive for efficient use and higher potential for waste. Price effectively becomes a measure of a resource's worth. When asymmetric or incomplete assessment of scarcity is prevalent, price may not properly indicate the cost of the resource being consumed.

In some areas of the world, forested land has been perceived as abundant and the resulting price for land has been limited to the perception of present period abundance. The net result of the perception has been excessive global deforestation, resulting in present period-pronounced scarcity in some regions. Decades will be required to promote regrowth of the same lands. Had prices considered the impact of forest harvesting, or the price of temporal scarcity, demand would have been lessened. Both consumption and production could have promoted efficient market pricing leading to sustainable resource use, all from this simple inclusion.

Demand and supply yield market outcomes that are assumed to represent an efficient allocation of resources. The price at which the quantity demanded equals the quantity supplied is therefore expected to embody the cost associated with the production and consumption of the good or service. However, production and consumption are not limited to the transactional nature of exchange of the final good at the determined market price. In the process of production and consumption there are costs that are not factored that impact the well being of the economy at large and these are referenced as externalities. In essence, externalities arise when an individual or firm engages in activities that influence the well being of others and where no compensation in provided in exchange for the imposition.

Typically externalities are characterized as negative, signifying that the externality yields an adverse outcome. These externalities are referenced as being *negative externalities*. However, there is a potential that a positive outcome could be generated leading to a positive externality. In the discussion of externalities it often assumed that market participants accept the externalities generated by their actions as acceptable due to their focus on immediate gratification of their needs. For the producer this equates to externalizing the cost of disposal of waste products into waterways and the air where no cost is directly borne to adversely impact profits but qualitative costs are assessed that may impact the enjoyment and longevity of multiple life forms and generations of human life. For the consumer the externality can be evaluated in the indifference to waste creation at the point of the consumption decision or even the externalities associated with the production of the good or service being purchased. In the case of the former, the cost of disposal of packaging material is typically marginal to zero, relatively negligible, but disposal creates a negative externality in the landfill, incinerator or recycling plant that could have been avoided with a thoughtful exercise of demand.

At present, the type of internalizing of externalities that has occurred has been limited to quantifying the externality to an overt cost. However, to the extent that the costs may remain unassessed and the market mechanism is not cognizant and focused on the elimination of the externality-based cost, rather the minimization of overall costs, this process has yielded suboptimal outcomes. For example, assume that a firm produces ambient pollution as a result of incineration of waste. If a governmental regulatory body institutes a fee or cost for pollution, effectively charging the firm for the ability to pollute the air, the producer is able to delegate responsibility for environmental stewardship to the price of pollution. Additionally, depending on the price elasticity of demand for the service offered, the producer may be able to not only transfer the costs now associated with polluting activity to the consumer, but may also be able maintain the pollution level. Assuming that the consumer is inelastic, in this example the negative externality related to internalizing the cost has not changed, instead only the responsibility of pollution has been transferred to a cost, revenue to the regulating body has been generated, and the consumer has suffered erosion in their overall disposable income and purchasing power.

The same type of scenario exists with a permit trading program, where in effect permits are issued for a specific amount of externality emission, allowing economic agents to trade and thereby optimize through again cost minimization. However, the cost minimization is founded on the presumption or delegation of the permit system to fostering socially optimal outcomes, again, relieving the economic agent engaged in the creation of the externality form being directly accountable for qualitative actions. Additionally, the trading of permits assumes that optimal financial outcomes equate to optimal environmental and social outcome due to the aggregated assessment of pollution. However, to the extent that pollution is not distributed evenly and certain locations may have a disproportionate concentration, the permit systems fails to generate a socially optimal outcome. This may be compounded by the impact of inelasticity, which may allow for the transfer of costs of implementation of the permit program to the economic agents the program was designed to protect.

Externalities are defined as a type of market failure based on the premise that optimal social outcomes result from individual economic agents acting in self-interest. However, if instead of being a market failure, externalities could be evaluated to assess and develop an optimizing strategy between individual interests and enhanced social outcomes, externalities could be internalized within the market model as a modification of preference. Perhaps externalities only indicate a lack of holistic awareness on the part of the consumer and producer or a cultural bias toward immediate gratification. These characteristics can be potentially modified through education. Optimal and universally acceptable strategies could then be adopted to promote sustainability.

The success of this internalization strategy relies on the development of the educated rational economic agent as a consumer. If consumers are aware of the responsibility inherent in their consumption and are aware of the environmental and social impact of production processes, consumer demand can create the coalescing framework to augment preference to exhibit demand for sustainably produced products. The augmentation in demand does not allow for the opportunity of delegation of responsibility of pollution capacity to a cost or alternatively, the incorporation within a cost minimization framework, as a result, the change in preference and subsequent modification in demand promotes the development of market outcomes that are environmentally and socially optimal from the position of what is supplied.

Resources such as air and water have no market price and are considered to be abundant. On the surface, these resources may appear to be unlimited however, increased population pressures along with externalized costs related to production, such as pollution, have diminished the availability of both potable water and clean air. How could this have occurred?

The lack of price, a market model promoting the focus of profit maximization, and promotion and validation of unlimited wants are largely responsible. Consumers have effectively allowed supply to determine demand by not imposing restrictions on how goods can be produced. Producers have focused on short-term profitability in lieu of long-term strategic resource utilization. In the short-run both consumers and producers have benefitted but the cost of consumption and profitability was externalized to other nations, the environment, and future generations. For example, in the seventeenth century, North American coastal waters were described and recorded as being rich in quantity and diversity of fish; the perception of abundance led over time to overfishing and presently many varieties are endangered or at the risk of extinction. The cost of fishing included the human and capital costs not the replenishment costs. This yielded an ability to maintain artificially low prices, greater yields for profitability (over fishing), and waste.

6 Market Prices, Values, and Common Goods

An understanding of the perception of scarcity and abundance provides a strong foundation to understanding supply, demand, and market outcomes as these concepts relate to resource allocation and sustainability. To the extent that consumers delegate responsibility for sustainable consumption to producers and producers are focused solely on profit maximization increased understanding of the responsibility inherent in consumption may provide a catalyst for increasing sustainable production, consumption and development. As holistic evaluation of consumption is an assumed behavior of the rational economic agent, strengthening the understanding of the role of consumption may be significant in enabling the development of the rational economic agent.

Supply and demand reflect the amount that producers or suppliers of a good or service are willing and able to sell at a particular price and the amount that consumers of a good or service are willing and able to purchase at a particular price, respectively. Though on the surface the concepts of supply and demand appear simple the characteristics that determine the explicit willingness and ability can be complex. The complications can arise as a result of differences in the preferences, behaviors, cultural values, financial capacity, as well as resource access and availability to the production process as these relate to suppliers. For consumers or demand, the complications can also be attributed to preferences, behaviors, cultural values, financial capacity and wealth perception, as well as the perception of value and price, along with access and availability, of other substitute and complementary goods. Where and how the supply and demand interact with each other define a market. A market is comprised of a group of producers (supply) and consumers (demand) for a specific good or service, who collectively, as part of their exchange process, determine the market price or equilibrium price of a good or service.

Price is the natural outcome of the supply and demand relationship. It is indicative of the value of a good based on a consumer's assessment of the costs and benefits of purchasing the good. As consumers become increasingly aware of the environmental and social costs of production, the prevailing price may be corrected either through regulatory imposition of the costs of externalities within the market mechanism or via consumers, who will opt to purchase goods not on price but related to holistic production costs.

It is important to note that the market relationship is dependent on information and understanding of the limits of duty of care. The outcome of the market relationship, price and quantity, can only reflect the embedded preferences and social values depicted in demand and supply. If the market outcome does not meet expectations, the market model is not to blame; rather the prevailing value structure may be the flaw.

Value in this context is related to how resources are valued from the perspective of the quality of care and maintenance we would be willing and able to provide to ensure the protection of the resource. The use of the word "value" is not directly based on market quantification but expresses the hierarchical importance that consumers and producers would attribute to a resource; examples may include the environment, human health, and animal welfare.

Every day consumers make decisions with the collective strength of aggregated individual demand. These decisions influence supply and demand going forward, including the ability of producers to develop new goods and services, as well as resources and technological advances to satisfy both existing demand and projected future demand. Demand is a powerful catalyst in the evolution of market outcomes. However, to a large extent the power of demand is limited both by the fragmentation of consumers due to limited opportunities for coalescing around specific interests and limited consumer understanding of the inherent power of aggregated consumption decisions. From this perspective, understanding how the market functions and the power of consumption in creating sustainable economic outcomes is one aspect of developing into a rational agent.

The values embedded and communicated within demand and supply, determine the manner in which a need or want is attained. To the extent that there is no discussion of the values and behavioral factors assumed and reflected in demand and supply, arguably, implicit values, the values and the subsequent behaviors become endogenous to the economic system. From this perspective, explicit awareness of present behavioral assumptions inclusive of the "unlimited wants" of consumers, profit maximization motivations of producers, and the understated resource depletion resulting from externalized costs, offer the potential to modify active and embedded behavior.

7 Conclusion: Conscious Consumption and the Social Norm of Sustainability

The explicit discussion of the embedded assumptions guiding the behavior of the decision-maker is typically not a part of the economic education process. As a result, to the extent that individual economic agents, producers or consumers of a good or service, are bounded by rationality that does not include addressing the impact of externalized or non-quantified costs, the economic discussion does not promote or position the assessment of alternative outcomes. Implicitly and endogenously, the economic discussion establishes and maintains consumption to production circular flow, focusing on the gratification of consumption and profit taking from production, seemingly eliminating assessment of externalities and holistic dynamics. Returning to the milk carton example provided in the **Introduction**, the economic discussion would be limited to the utility gained from consuming the milk and the corresponding profit maximization of the producer. Waste would be regarded as an externality rather than an endogenous aspect of the decision making process. Additionally, costs are priced into the product through efficient market assumptions. In net consumers would expect that the purchase price is indicative of the holistic cost of the product and producers would view production costs as being related to market priced inputs not environmental impacts during or as part of the life cycle of the good.

Economics, in present practice, evaluates efficiency with respect to the use of resources to maximize production and consumption, not by the moral desirability of the physical methods and social institutions used to achieve this end. The factors that are included in an economic evaluation are limited to the tangible quantifiable costs and costs are overlooked where either a market or regulatory oversight has not provided a monetary justification (Shah 1999). From this perspective, the impact of consumption decisions on the environment, economic disparity, or endangerment of other species is not an issue. The market mechanism disenfranchises the consumer from the welfare of those impacted by his/her consumption and promotes the perception that price alone is indicative of the true cost of a good. The possibility that consumption should be reduced because the act of consumption is not good for the soul, or is not what actually makes people happy, has no place within the economic value system (Nelson 1995). The underlying assumption is that consumers are driven to want more. As a result, economic modeling assumes that reduction in consumption in the current period is only addressed through the lens of an increase in consumption in a later period (Knoedler and Underwood 2003). That the assumption of insatiable want may be taught a learned behavior, reinforced through a market model is not even addressed in economics.

A general and seemingly applicable assumption is that consumers and producers maximize the benefit related to the opportunity accessible in their particular circumstance. The desire to reach an optimal outcome for a given point in time, as has been noted before, is subjective and specific to how these economic agents view the concept of maximization, which in turn is likely to be highly correlated with cultural values. For example, in Indigenous societies there is evidence that a balance between present and future periods along with that of the environmental system, as a whole, was included in decision-making and optimization (Nerburn 1999). In present consumerism fostered economies, the cultural values are less likely or unlikely to incorporate environmental and social justice parameters proactively. The focus of observable and marketed consumption is immediate gratification. However, as consumer awareness of both the impact of consumption and the power of consumption to modify and catalyze economic outcomes increases there is growing evidence of a shifting cultural paradigm to one of sustainability.

Markets do fail to produce optimal outcomes. Sometimes this is due to the myopic focus of market participants as in the case of externalities and in other circumstances it can be attributable to the lack of excludability as in the case of common goods. To some extent cultural values dictate the significance of the adversity related to the creation of externalities or abuse of common goods. The use of market models has been the regulatory mechanism to modify socially non-optimal outcomes, but through relying on the market mechanism rather than simultaneously including mechanics to promote cultural change, the majority of regulatory interventions to date have had limited to questionable success.

A constituency with an understanding of the holistic relationship between consumption and sustainability and having engagement in government are foundational elements in achieving and maintaining sustainability as a cultural norm. For long-term traction, sustainability is dependent upon holistic and routine evaluation of economic and societal frameworks. These frameworks need to be assessed and modified as part of an on-going continuous improvement process. Fundamentally, what may have been viewed as appropriate action at a point in time may no longer serve the same purpose due to changing environmental, social and cultural parameters. However, the members of a society have to be both empowered and cognizant of the need for this type of evaluation in order for efficiency and ultimately sustainability to be a realized inter- and intra-generational attribute. From this perspective, the deployment of consumer education programs targeted at defining responsible demand, conscious consumption, are a requisite foundation for sustainability (Junyent and de Ciurana 2008).

References

Bentham, J. (1879). *An introduction to the principles of morals and legislation* (p. 14). Oxford: Clarendon.

Boran, I. (2006). Benefits, intentions, and the principle of fairness. *Canadian Journal of Philosophy, 36*(1), 95–115.

Colander, D. (2005). What economists teach and what economists do. *The Journal of Economic Education, 36*(3), 249–260.

Czech, B. (2000). Economic growth as the limiting factor for wildlife conservation. *Wildlife Society Bulletin, 28*(1), 4–15.

Day, G. S., & Aaker, D. A. (1970). A guide to consumerism. *Journal of Marketing, 34*(3), 12–19.

Fagg, J. (1981). The fundamental principles of economics. *Journal of Economic Issues, 15*(4), 937–942.

Junyent, M., & Geli de Ciurana, A. M. (2008). Education for sustainability in university studies: A model for reorienting the curriculum. *British Educational Research Journal, 34*(6), 763–782.

Knoedler, J. T., & Underwood, D. A. (2003). Teaching the principles of economics: A proposal for a multi-paradigmatic approach. *Journal of Economic Issues, 37*(3), 697–725.

O'Hara, S. U. (1995). Sustainability: Social and ecological dimensions. *Review of Social Economy, 53*(4), 529–551.

Mill, J. S. (2016). In W. J. Ashley (Ed.), *Principles of Political Economy with some of their Applications to Social Philosophy* (7th ed.). London: Longmans, Green and Co. (1909).

Marx, K. (1968) 28. In *Das Kapital* (Vol. 1).

Nelson, R. H. (1995). Sustainability, efficiency, and God: Economic values and the sustainability debate. *Annual Review of Ecology and Systematics, 26,* 135–154.
Nerburn, K. (1999). *Wisdom of the native Americans* (p. 41). Novato, CA: New World Library.
Ricardo, D. (1911). *The principles of political economy & taxation* (p. 8). London: J.M. Dent & Sons.
Schweitzer, A. (1981). Social values in economics. *Review of Social Economy, 39*(3), 257–278.
Shah, M. (1999). Synthesis of ecology and economics: Towards a new theoretical paradigm. *Economic and Political Weekly, 34*(46/47), 3293–3298.
Smith, A. (1791). *An inquiry into the nature and causes of the wealth of nations* (p. 22). Basil: Tourneisen and Legrand.
UNESCO (Culture Sector). (2000). *World culture report.* Paris: United Nations.
Venkatesan, M. (2015). Values, behaviors and economic outcomes. Retrieved from https://www.youtube.com/watch?v=ipq5owxlU0g.
Venkatesan, M. (2016). *Economic principles: A primer, a foundation in sustainable practices.* Mathews, NC: Kona Publishing.

Greening Networks: Mapping Sustainability Beyond Institutional Boundaries

Fayyaz Vellani and Naveed Nanjee

Abstract This chapter examines the incorporation of sustainability into higher education curricula through a variety of modes, and considers the challenges of assessing this still newly evolving field. Such an undertaking is necessarily complex, as it is not standard practice for universities to mandate curricular changes in general, or those specifically related to sustainability. Previous studies have highlighted the value of incorporating sustainability into postsecondary curricula, mostly focusing on how these processes operate at the institutional level. We argue that a broader lens of analysis is required, including the role of consortia of non-governmental organizations and researchers, professional academic associations, and collaborations between students, staff and faculty. The chapter uses case studies focusing on these various networks in order to better understand these efforts.

1 Introduction

This book is written from the premise that incorporating sustainability into higher education curricula is a desirable and necessary goal. In this chapter, we argue that the structured and at times hierarchical nature of universities presents both challenges and opportunities, hence the impetus for creating networks which can nimbly circumvent and transcend institutional boundaries. Reflections on how the efforts of those who are committed to including sustainability in university curricula can be optimized must include innovative methods for the dissemination of best practices, strategies for overcoming institutional resistance, and ways of incorporating the latest knowledge into these efforts. Given the continually evolving nature

F. Vellani (✉)
Critical Writing Program, University of Pennsylvania, 3808 Walnut Street, Philadelphia, PA 19104, USA
e-mail: fayyaz@writing.upenn.edu

N. Nanjee
University of Edinburgh, 54/1 Rankeillor Street, Edinburgh EH8 9HZ, UK

J.P. Davim (ed.), *Curricula for Sustainability in Higher Education*, Management and Industrial Engineering, DOI 10.1007/978-3-319-56505-7_3

of our understanding of environmental issues, those engaged in this work must necessarily strive to be on the vanguard of knowledge in this area.

While the notion that incorporating sustainability into universities' work is a relatively recent one, it is quickly gaining acceptance, as evidenced by the existence of publications such as the *International Journal of Sustainability in Higher Education*. Significant research has been conducted at the disciplinary level (see Sukumaran et al. 2004 for example), the conceptual level (Filho 2000), as well as at the global level (Blewitt and Cullingford 2013). In spite of such studies' findings, it is challenging to assess at what rate universities are actually addressing sustainability, and with what levels of success. This is partly due to the dearth of studies in this area. Additionally, it is challenging to measure and assess what actually occurs in classrooms, for they are spaces which—though public in a certain sense—remain the domain of individual instructors.

As a starting point for sustainability action, it is both necessary and important to develop laws and policies. Indeed, there have been several of these in the United States beginning in 2007: the Energy Sustainability and Efficiency Grants and Loans as part of the Energy Independence and Security Act,[1] followed by The University Sustainability Program in 2008 in conjunction with the Higher Education Opportunity Act.[2] At the sector level, the Association for the Advancement of Sustainability in Higher Education (AASHE) works to promote good environmental practice among providers of higher education across the USA. Foremost among its wide range of initiatives is the recognition of post-secondary institutions which have achieved demonstrable success with so-called green initiatives both on their campuses and beyond. AASHE evaluates these efforts using a Sustainability Tracking, Assessment and Rating System (STARS), "a voluntary system that allows different colleges and universities to report trends and track their sustainability efforts."[3]

Despite these extensive efforts, law and policy are not a panacea. Even where such legal initiatives exist, they can be undermined by a lack of enforcement, funding, and other necessary resources. For example, within a year of the enactment of the University Sustainability Program, members of the U.S. Congress were calling for it to be funded at the levels originally envisaged in the legislation, as the program had not been adequately funded even after the law came into effect.[4] Furthermore, the notion that governments and other governing bodies alone are responsible for enacting change and undertaking efforts to improve environmental conditions is insufficient. It is particularly dangerous for those committed to sustainability to simply expect governments to act in ways which are environmentally ethical, given the shorter-term perspectives of elected officials who think in two-year (U.S. House of Representatives), four-year (U.S. Presidency), five-year

[1]https://www.epa.gov/laws-regulations/summary-energy-independence-and-security-act.

[2]http://www2.ed.gov/policy/highered/leg/hea08/index.html.

[3]http://www.bestcolleges.com/features/greenest-universities/.

[4]http://www.fundee.org/campaigns/usp/.

(U.K. and Canadian parliaments) or at most six-year cycles (U.S. Senate). This is what Mark Carney, the Governor of the Bank of England, called the Tragedy of the Horizon in a 2015 speech. Furthermore, even where groundbreaking legislation, such as the U.S. Clean Air Act (1970), does exist, conditions change, political environments change, and most critically, companies find ways of working around such laws while remaining technically in compliance.

In the face of these challenges, those committed to sustainability in universities —namely researchers, professors, administrators, and students—must undertake significant responsibility. Given all that has happened even since the establishment of the Environmental Protection Agency (EPA), it would be prudent to not entrust the responsibility of enacting sustainability work entirely to governments or private corporations. This is a not polemical claim about the morality of these actors: it is unsurprising that politicians should be inherently guided by their quest for votes (and increasingly, donors), while private sector corporations are primarily motivated by profit, which is built into their very structure and purpose. It therefore falls to those in the world of higher education, who are guided by the pursuit of knowledge and excellence, and whose institutions' primary aims revolve around the important work of educating future generations of leaders.

Given the important role to be played by universities, it is not enough to hope that all institutions will simply adopt policies aimed at saving the planet from environmental destruction. An undertaking of such magnitude requires the engagement of actors at all levels, and universities are particularly well placed—given their ever-expanding pursuits of new information—to lead the charge. It follows that new, qualitative metrics are needed in order to ascertain the depth of sustainability action taking place at universities. In particular, this chapter examines modes of collaboration which go beyond conventional boundaries, and elucidate relationships and networks of power-knowledge which transcend our conventional notions of these institutional actors (Foucault 1978).

2 Literature Review

Recent geopolitical events have caused consternation among some in the environmental field. Both the UK's vote to leave the European Union and the 2016 U.S. Presidential election have raised the specter of stalled progress on what had hitherto been seen as steps in the right direction. For example, in the second term of the presidency of Barack Obama, the U.S. government underwent a 75-point Climate Action Plan, directing the federal government to begin planning for climate change. President Obama enacted a series of executive actions in lieu of working with the U.S. Congress, including reaching an agreement with China to reduce carbon pollution in both countries, as well as the Clean Power Plan, relying on the EPA to enforce the reduction of carbon dioxide emissions from power plants by 32% by 2030. Perhaps most significantly, at the conference of the parties (COP 21) in December 2015, 197 countries signed the United Nations Framework on Climate

Change's Paris Agreement (UNFCCC). When asked directly about whether or not it is possible for the incoming president to rescind the United States' signature to the agreement (notwithstanding its lack of ratification), President Obama answered:

> Well, historically what happens is that when you have an international agreement, it carries over into the next administration. There were agreements that President Bush made that I respected, because as president of the United States, it was important for me to project a sense of continuity in the U.S. government…The good news is that a lot of these initiatives that we've taken work, and don't just work in terms of reducing emissions, they work from an economic perspective. And so over the course of my eight years, when we doubled clean-energy production or we cut [auto emissions] in half, that wasn't just a matter of regulations that can suddenly be erased; that had to do with investors and businesses and utilities and consumers all organizing themselves, figuring out that, you know what, being smart on energy is good for the planet and it's good for my pocketbook.[5]

When pushed by the interviewer, Wenner (2016) with the question: "I understand all that, but you have nearly all of the science saying we are past the tipping point, and you've got the Koch brothers financing an absolutely obstructionist Congress. That's not going to change. Their ideology seems to be set on the subject. The money that's bought these votes is set on the subject", Obama responded:

> Yeah, listen. If you want to persuade me that everything is going to be terrible, then we can talk ourselves into that. Or we can act. It is what it is. There's been an election. There's going to be a Trump presidency, and Republicans are going to control Congress. And the question is going to be, for those like you and I, who care about these issues, do we figure out how to continue to make progress in this environment until we have a chance for the next election. And will we have mobilized ourselves and persuaded enough people that we can get back on a path that we think is going to be helpful for families, helpful for the environment, helpful for our safety and security and rule of law and civil rights and social rights?

The French Philosopher Michel Foucault was among the first thinkers to challenge traditional notions of power: that it is necessarily coercive or authoritarian, or that it always rests among only the sovereign, the monarch, or the ruler. To Foucault, power is dispersed, pervasive, ordinary; neither top-down nor bottom up, but diffuse. As he notes: "Power is everywhere; not because it embraces everything, but because it comes from everywhere" (1978: 93). This is an important principle for scholars and researchers committed to sustainability. The notion of Actor-Network theory, first promulgated by Callon (1991), Latour (1992) and others, posits that great scientific strides are made not due to heroic individuals, but rather owing to the existence of discrete, varied and multifaceted networks. Proponents of Actor-Network Theory theorize a principle of generalized symmetry, namely that "what is human and non-human (e.g. artifacts, organization structures) should be integrated into the same conceptual framework and assigned equal amounts of agency."[6]

---

[5]Wenner (2016).

[6]https://www.learning-theories.com/actor-network-theory-ant.html.

If, as Foucault (1978) notes, power is not static but rather in a state of flux, then we all have agency in claiming and acting upon the power we possess. Foucault contests grand notions of truth, instead noting that the prevalent ideas to which many agree are the results of scientific discourse and the work of institutions, which are themselves contested, and can be continually altered through educational systems. Taking up Foucault's mantle, Hayward writes about power in the public education system in the USA, calling for a "de-facing" of power, she argues that those committed to making social change should focus on "whether the social boundaries defining key practices and institutions produce entrenched differences in the field of what is possible" (1998: 20).

With this focus on highlighting boundaries with a view to surpassing them, the case studies in this chapter examine collaborations: (1) between researchers and those outside of academic institutions, (2) among academics within particular disciplines, and (3) between institutions and their students. Our analysis therefore takes us not only into institutions, but also attempts to look both around and outside of them, as well as seeking to understand the efforts being undertaken to overcome what some see as the division of power between administrators, professors and students. Indeed, the chapter ends with a focus on the role of students themselves in shaping not only the institutions to which they belong, but also the broader domain of knowledge.

3 Case Study One: Consortia of Researchers and Non-governmental Organizations

In examining the myriad individuals and groups—both inside and outside of universities—who collaborate on various initiatives for sustainability, the aim is to understand how these actors come together at the ideation stage. This means not only producing scientific papers and reports, but also efforts for sustainability which can translate effectively into universities' priorities. While knowledge in such fields as climate science has steadily advanced, these new domains have not necessarily found their way into either universities' curricula or institutional plans more broadly, which is ironic given that these institutions are often the site of such advances. As Dominelli (1997: viii) notes, "it has been said that the teacher today is teaching what has been learned at least two generations ago from textbooks that were written several decades before."

Over the last few decades, an increasing number of researchers have been working to translate their findings into tangible action outside of academia. Too often, the lag time in curricular change—as noted in the quote by Dominelli above—as well as institutional boundaries and barriers, means that this research on the vanguard gets published in eminent journals, but potentially not acted upon when those same universities who foster this research take decisions affecting sustainability. This has been evidenced most recently with the contentious issue of divestment, for example, wherein a number of prominent institutions have decided against divesting their institutional funds from companies involved in the fossil fuel industry (see Vellani and Nanjee 2016, for example). This section of the chapter therefore explores the nature of such groupings of NGOs and considers questions about their connection to university curricula.

Since the Stockholm Conference in 1972, universities have participated in the development and implementation of sustainable development through networks, charters, declarations, and conferences. The UN Decade of Education for Sustainable Development (2005–2014) spearheaded by UNESCO, brought universities together, and various other stakeholders, to mobilize efforts for a sustainable future through education, research, operations, or outreach. Camarinha-Matos et al. (2010) suggest that challenges facing sustainability need collaboration in order to make the necessary changes required. These changes are often much greater than the capability of individual actors. Thus, collaborative networks have a critical role as a "multi-stakeholder collaborative perspective" is needed to further efforts towards sustainability. Such efforts have played a critical role in establishing, curating, and furthering collaborative networks over the course of past few decades by actors working with a wide range of stakeholders to address issues of sustainability in various fields of research as well as through policy.

Globally, at an institutional level, there have been a number of university alliances and networks which have promulgated the importance of higher education institutions to play a major role to address the challenges of sustainability. The *Talloires Declaration* in 1990 called for university leaders to take initiative and mobilize resources to address environmental degradation (ULSF 1990). The *Talloires Declaration,* signed initially by twenty university presidents and vice chancellors from around the world, was the first declaration of its kind specifically for institutions of higher learning and provided a model for universities to collaborate and make commitments around subsequent declarations (Lozano et al. 2013).

In 1993, the International Association of Universities (IAU), which included 1200 member institutions at the time, adopted the *Kyoto Declaration on Sustainable Development*, which urged universities "to teach and undertake research and action in society in sustainable development principles, to increase environmental literacy, and to enhance the understanding of environmental ethics within the university and with the public at large" (IAU 1993, para 4). The University Charter for Sustainable Development, an initiative of the Co-operation Programme in Europe for Research on Nature and Industry through the Coordinated University Studies (COPERNICUS), further elaborated to role that sustainability should play in curriculum, as it outlined:

> Universities shall incorporate an environmental perspective in all their work and set up environmental education programmes involving both teachers and researchers as well as students - all of whom should be exposed to the global challenges of environment and development, irrespective of their field of stud… Universities shall encourage interdisciplinary and collaborative education and research programmes related to sustainable development as part of the institution's central mission (COPERNICUS 1994).

To date, over 320 European Universities have signed the University Charter. Furthermore, the *Bonn Declaration* signed at the UNESCO World Conference on Education for Sustainable Development in 2009 built on the previous work and at a practice level urged:

> Support the incorporation of sustainable development issues using an integrated and systemic approach in formal education as well as in non-formal and informal education at all levels, in particular through the development of effective pedagogical approaches, teacher education, teaching practice, curricula, learning materials, and education leadership development, and also by recognizing the significant contribution of non-formal education and informal learning as well as vocational and workplace learning. Sustainable development is a cross-cutting theme with relevance to all disciplines and sectors (UNESCO 2009).

The Nagoya Declaration on Higher Education for Sustainable Development (Nagoya Declaration) was adopted by the participants of the International Conference on Higher Education for Sustainable Development in Nagoya, Japan on November 9, 2014, which confirms the responsibility of the higher education sector for pursuing sustainable development.

The *Nagoya Declaration* highlights the importance for higher education institutions' commitment to sustainable development, which is also supported by the UNFCCC recognizing universities and institutions of higher education as part one of the nine civil society constituencies. It reaffirmed higher education institutions' involvement in sustainable development as these institutions play a critical role in the shaping of the global development agenda.

Leading up to the Paris Agreement, an alliance of universities, institutions, and students came together to write an open letter to the president of COP 21, confirming their commitment to implementing solutions to mitigate climate change. This collaborative network was inspired by the Higher Education Sustainability Initiative (HESI). HESI, established prior to the United Nations Conference on Sustainable Development in Rio de Janeiro (Rio+20), has a membership of over 300 universities from around the world. The commitments made by universities part of HESI accounted for more than one-third of all the voluntary commitments from Rio+20.

Several authors suggest that the Paris Agreement would not have been signed without the collaboration and work of over one thousand non-governmental organizations (NGOs) who were part of the negotiations at COP 21 (Jacobs 2016). Since first COP 1 in Berlin the number of civil society organizations and specifically higher education institutions participating in international climate change conference have been dramatically increasing, which signifies the importance of the negotiations for various stakeholders. By participating in the process, civil society has had an increasing influence on the outcomes of negotiations, as it is also able to obtain information first-hand throughout the process. Ban Ki Moon, the United Nations Secretary General, offered his thanks to civil society at the end of the negotiations at COP 21, stating, "We would not have achieved this victory without the leadership and vision of civil society. You have highlighted both the stakes and the solutions to the climate challenge. From young activists to artists, faith leaders to business leaders, mayors to mothers, all of society has come together under one banner and brought forth this moment of hope" (Ki-Moon 2015).

Universities played a critical role in bringing these vast collaborative networks together, which made a significant contribution to the adoption of the Paris Agreement. This demonstrated the power of networks that were able to break down traditional institutional barriers which had prevented such agendas from being adopted for several decades. Networks like HESI, COPERNICUS, and IAU created pathways towards potential solutions in sustainability. The multitude of declarations, charters and conferences have cultivated a consensus of leadership within the institutions of higher education. However, the adoption and implementations of these declarations into universities' curricula, research, and outreach at many institutions have yet to impact the classroom. The challenge now is for these institutions and networks to help with the implementation of these policies in order to impel progress beyond the ideation stage.

While it is clear that academic research feeds into this important policy work, do these deliberations make their way into university classrooms? In what ways might we measure and assess this translation into teaching? Apart from those researchers engaged in such collaborative intellectual efforts discussing this work in the classroom, there is very little to suggest that research on the vanguard is making its way into higher education curricula. Perhaps the most well-known model of this combination of research and teaching in the field of environmental studies is Columbia University's Earth Institute, which also benefits from its location in a city with many prominent sites of international collaboration, most notably the United Nations. What seems to be a key factor—particularly at research oriented institutions—is that those who are most focused on research are rewarded for their outputs with teaching relief. In some cases, this is further incentivized by a funding system in which grants are awarded to "buy out" a professor's teaching time. There remains potential for teaching and research to be more complementary and further integrated. The current career trajectory for academics, which is increasingly focused on the publication of research in peer-reviewed journals, does suggest a structured manner which would readily allow for this cutting edge research conducted outside of universities to be included in university curricula.

4 Case Study Two: Professional Associations

This section of the chapter examines the nature of the work being undertaken at disciplinary and professional levels to incorporate sustainability into curricula across a range of academic fields of study. National professional bodies represent an important tool which can aid in the task of gleaning how universities are making sustainability part of their most important work: teaching. While almost all universities worldwide have published policy statements detailing their commitments to sustainability, the national professional bodies we highlight are useful because of their roles in guiding the scholarly discourses within academic fields of study. In some cases, these discourses are related to how universities are recognized and/or accredited within their respective fields. Additionally, these professional bodies are linking their work to broader policy discussions, which can be useful for framing and highlighting the work of university teachers at the national level, and creating two-way dialogues which can be informative for both governments and academics.

Professional bodies in respective fields of higher education (both accrediting and non-accrediting discipline-based organizations) represent opportunities for implementing changes to curricula which go beyond the boundaries of a particular universities. Those working in academia are known to collaborate extensively; we here suggest that this kind of intra-disciplinary examination of practices related to sustainability ought to be taken up by the myriad professional associations which exist, both within the USA and beyond. For example, in 2009, the American Association of Colleges of Nursing (AACN) published a document setting out its vision, entitled "Toward an Environmentally Sustainable Academic Enterprise": An AACN Guide for Nursing Education."

This collaboration originated in response to a White House initiative on public health at which Deans of Nursing Schools from across the USA had gathered. The challenge of addressing sustainability in their curricula was then put to these Deans, and the AACN's subsequent work on this initiative culminated in the aforementioned report and guide. The main outcomes suggested in this document were to: (1) identify best practices related to environmental sustainability and make recommendations for their dissemination, and (2) recommend content and competencies which would aim to make the curricula sustainable for both undergraduate and graduate education in the field of nursing. While the AACN represents nursing schools at more than 765 institutions of higher education in the United States, it is one of a large number of school accreditors, and even larger than this number are those associations that are not recognized as accreditors. From a sustainability perspective, these associations maintain significant influence over how their subject matter is taught, regardless of whether or not the association is an accrediting body.

The exercise undertaken by the AACN demonstrates that such guidance can be created not only by accrediting bodies and/or professional associations with a purview over their given fields of study (e.g. in the professions of nursing, medicine, accounting, engineering, and psychology), but also in discipline-based organizations such as the Association of American Geographers (AAG). For example, in 2012 the AAG collaborated with the *Journal of Sustainability Education* to create a special issue focused on geography which included articles on the role of geography in teaching students about sustainability. As the discipline of geography is fundamentally concerned with questions of scale, it is an ideal site for pursuing questions about what scale of activity humans should be undertaking in the pursuit of sustaining life on the planet. In the early 2000s, the AAG undertook a project entitled Global Change in Local Places which examined the "scale of global climate change, impacts, and adaptation."[7] One of the key findings of this project was that the extent of global knowledge of sustainability was not well-matched with local actions.

[7]http://www.aag.org/cs/annualmeeting/scale.

More than a decade later, in 2014, as part of its annual meeting, the association devised a theme related to Scale and Sustainability which considered sustainability from a holistic perspective, with a view to balancing the environment and human development without compromising the needs of either. Philosophically, this evokes the spirit of the 1987 Brundtland Report by the World Commission on Environment and Development. The AAG sessions organized around the Scale and Sustainability theme at the 2014 annual meeting largely focused on how sustainability goals could be implemented temporally and spatially. While the 2014 meeting included 24 sessions organized around this theme, none appeared to address the application of sustainability to the higher education sector in general, or specifically to curricula. The AAG is, on the whole, deeply committed to exploring knowledge of how humans can enhance their sustainability, not only from a theoretical perspective, but also in more pragmatic ways. For example, as its 2016 annual meeting was held in San Francisco, many local projects, which meeting attendees were invited to visit, revolved around environmental gentrification, a theme of particular relevance in that city.

It is admirable and essential that both the AACN and the AAG are undertaking the work of making their profession and discipline, respectively, more accountable for practices whose explicitly stated goal is sustainability. The work of these two associations came about for particular reasons. For the AACN, it was partly in response to the Affordable Care Act, and the challenge set to the nursing profession by the White House to try and weave sustainability into their everyday practices. For the AAG, environmental questions are integral to the discipline of geography, concerned as it is with "the study of places and the relationships between people and their environments."[8]

Beyond these two exemplars, the question still remains: in what ways can professional and disciplinary associations equip academics to incorporate sustainability into their teaching? Furthermore, if and when this occurs, how can such work measured and assessed for its effectiveness? There is yet much more research to be conducted in order to overcome institutional and disciplinary boundaries which can act as impediments to change. As bell hooks notes, "I'm really into the deinstitutionalization of learning and of experience. The more I've been in the academy, the more I think about Foucault's *Discipline and Punish: The Birth of the Prison* and the whole idea of how institutions work" (1994: 232). We are still at the developmental stages of the important work of fostering collaboration for sustainability within academic disciplines and professions.

5 Case Study Three: Student-Staff Collaboration

This final section of the chapter examines the role that students play in their own education, specifically in building sustainability into the higher education institutions

[8]http://nationalgeographic.org/education/what-is-geography/.

they attend. Student agency, voice and experience are increasingly important concerns for those engaged in university education. Writers in the *Chronicle of Higher Education* and other salient publications are increasingly interested in representations of students as active rather than passive learners, and as needing to be cognizant of the ethical implications of their actions.[9] At the same time, others have argued that simply asking for students' opinions on all matters, in an unstructured manner, can be counterproductive.[10]

In the area of sustainability, much of the student engagement to date has been voluntary. While the role of student participation and activism is important, perhaps reaching its zenith with the 1968 riots in Paris, it cannot be the sole means of inspiring universities to make their practices more sustainable.[11] The Green Office Movement is a response to this need for structured ways in which students can contribute to this work. The Green Office Movement emerged out of the awarding of the UNESCO-Japan prize for Education for Sustainability to the University of Maastricht. The University of Maastricht founded its Green Office in 2010 with the aim of being "a student-led and staff-supported sustainability hub empowering students to drive sustainability transitions" (Maastricht University Green Office 2016).

Following in Maastricht University's footsteps, 14 other universities in 4 countries have opened Green Offices in Utrecht, Wageningen, Rotterdam, Groningen, Amsterdam, Velp, Eindhoven, Exeter, Greenwich, Canterbury, Konstanz, Berlin and Ghent. The Green Office Movement has a stated goal of engaging 10,000 students and staff in 100 Green Offices by the year 2025. To date, more than 2000 students have become involved in these efforts. One of the key features of such collaborative models is that they enable a university's organizational structure to remain attuned to student concerns, ideas and engagement. For example, in its summary report for 2015, the Green Office noted that "it can be observed that students are not sufficiently aware of sustainability initiatives, such as UMGO."[12] Because of the formalized structure of the Green Office, the university is able to redouble its efforts so as to improve awareness of its sustainability initiatives, including the addition of sustainability courses, more professorships, and research related to sustainability. While the Green Office lauds developments in the areas of the University's energy efficiency, staff commuting, catering and e-waste, the UMGO notes that its momentum can at times be stalled due to "institutional inertia, a lack of clear sustainability management structures beyond and integrated UMGO and extreme decentralisation."[13]

The Green Office model is holistic enough that it covers various aspects of university operations and governance, so that when it confronts institutional or

[9]http://www.chronicle.com/blogs/next/2012/05/01/did-anyone-ask-the-students-part-i/.

[10]http://opinionator.blogs.nytimes.com/2010/06/28/student-evaluations-part-two/?_r=0.

[11]http://www.nytimes.com/2008/05/11/world/europe/11iht-paris.4.12777919.html.

[12]http://greenofficemaastricht.nl/impact/overview-and-full-reports/.

[13]http://greenofficemaastricht.nl/impact/overview-and-full-reports/.

attitudinal barriers, these can be overcome. Specifically, the GO model includes consideration for sustainability in the following areas: efforts to reduce negative environmental impacts in the area of university operations, a commitment for the organization "to learn about its own sustainability performance",[14] including collaborations between members of staff and students, to enhance awareness about sustainable behavior throughout the university community (among students, staff and faculty), to integrate sustainability into the curriculum alongside capacity building for students in this area, and to "increase knowledge exchange between researchers across faculties and to support and enhance opportunities for student research on sustainability-related questions."[15] Maastricht University's Green Office Model locates students at the core of its mission, as a

> a student-driven university department that is responsible for managing Maastricht University's sustainability portfolio. Green Office initiates and coordinates sustainability projects at Maastricht University by empowering students and staff members (Maastricht University Green Office 2016).

The mission and efforts of the UMGO are rooted in the value of student engagement. Given this unique approach, one wonders in what ways might such a model of collaboration be replicated. The Green Office model was designed to be adaptive to its own institutional context and adaptable to other institutional contexts. At its core are open-source principles which are available to be "adapted to the context of each institution" (rootAbility 2016). Specifically, the Green Office Principles are based on the following core elements: (a) Students and Staff—a consortium comprised of individuals from both of these groups form the core of the Green Office, (b) Mandate—a University's Green Office requires a mandate from its institutional governance structures in order to "drive the sustainability transition of the university, by creating new impulses, connecting and empowering actors, improving communications or developing sustainability strategies." (rootAbility 2016), (c) Resources—the Green Office requires a budget in order to cover the cost of staff salaries, office space, training and the funding of various other projects, (d) Integration—It is vital that a Green Office, when formed, be incorporated into a university's organizational structure and is governed by an authoritative decision-making body. Members of the Green Office's staff can and should, in turn, serve on various sustainability committees throughout an institution, ensuring a continuity of dialogues and holistic approaches to making institutional practices more sustainable, (e) Collaboration—the Green Office works closely with similarly-minded actors both within and outside of institutions. To date, a strong network of Green Offices has been formed in Western Europe, (f) Training—to date, Green Offices have received support and training from rootAbility, a youth organization aimed at inspiring students to personally make the institutions they attend "more sustainable, resilient and fair" (rootAbility 2016).

---

[14]http://greenofficemaastricht.nl/about/.

[15]http://greenofficemaastricht.nl/about/.

While all of this work is laudable, and the model offers a clearly-stated set of principles, as well as a solid organizational structure, an examination of its geographical reach suggests it has greater potential to be taken up than has occurred thus far. While the model has only been formally replicated in Europe, universities in other jurisdictions are starting to acknowledge the benefits of fostering partnerships between members of staff engaged in sustainability work, and students who are passionate about this topic. For example, the University of Pennsylvania's Green Campus Partnership—which serves as the institution's office for sustainability initiatives—recently launched a Staff and Faculty Eco-Reps program to match its Student program of the same name.

6 Conclusions

The above three typologies of collaboration have indicated efforts which are being undertaken in order to break down barriers to sustainability. This is particularly salient to those outside of academic institutions, especially with regard to the crucial agreements to emerge from Paris in December 2015. These efforts to encourage academics to take up sustainability work is also occurring across professions such as Nursing and in disciplines such as Geography. The AACN and the AAG have created mechanisms for introducing notions of sustainability into curricula using existing discourses about norms, standards and ethics, for example, but there remains scope for more of such work to be conducted beyond these professional associations. Finally, a plethora of universities, primarily in Western Europe, and inspired by Maastricht University's Green Office model, are working to transcend divisions between members of the faculty, staff and students so that they can catalyze change within their universities. Central to this model is the supposition that students need to be engaged in the process of their own learning, and in the work of creating a sustainable world.

In light of the sociopolitical climate of 2016—in which a desire among portions of the electorate in the UK and the USA to return to an imagined idyllic period of mass industrialization—those who are committed to sustainability need to introduce the idea of its importance not only into universities, but also into wider discourses about the economy and national and global priorities. President Obama recently noted that the results of the 2016 presidential election do not necessarily spell disaster for environmental issues: "So this notion somehow that these irreversible tides have been unleashed, I think, surrenders our agency" (quoted in Remnick 2016). Indeed, agency—and responsibility—lies with us all. As one writer recently noted:

> That future is possible. It might even be probable. But it's not inevitable. We can choose to see climate change, and we can choose to do this before it's too late. So how can we escape the quagmire of denial? As it turns out, the first step isn't that hard: just talk about it. To your friends, your family, colleagues…By talking about it, we'll unlock solutions.

> And crucially, it's by talking about climate change that we'll break the silence that allows it to go unnoticed and ignored.[16]

Universities in particular, as sites of knowledge production, should foster such dialogue. It might be the case that professors engaged in research on environmental issues which could be complementary are unaware of each other, or too busy working in silos, producing excellent research but not yet overcoming institutional and professional barriers to cross-disciplinary dialogue. It could also be the case that students—for millennials and current generations of youth are highly aware of environmental issues—have much to offer by way of thoughts, questions and ideas. Engaging in dialogue, questioning existing assumptions, and working together, these students and professors have the potential to make tremendous progress on issues of sustainability, as Maastricht University's Green Office model has indicated. Finally, it remains for those inside universities—which after all, remain enclaves of power and privilege—to disseminate their work broadly outside of their institutions, and outside of academe, for the benefit of wider society and the planet, thus fulfilling education's ultimate goal and promise.

References

Blewitt, J., & Cullingford, C. (2013). *The sustainability curriculum: The challenge for higher education*. Chicago: Earthscan.

Callon, M. (1991). Techno-economic networks and irreversibility. In J. Law (Ed.), *A sociology of monsters? Essays on power, technology and domination, sociological review* (Vol. 38, pp. 132–161).

Camarinha-Matos, L. M., Afsarmanesh, H., & Boucher, X. (2010). The role of collaborative networks in sustainability. In *Working Conference on Virtual Enterprises* (pp. 1–16). Berlin Heidelberg: Springer.

Carney, M. (2015). Breaking the tragedy of the horizon—climate change and financial stability—speech by Mark Carney. Retrieved November 01, 2016, from http://www.bankofengland.co.uk/publications/Pages/speeches/2015/844.aspx.

COPERNICUS. (1994). The University charter for sustainable development. Retrieved November 22, 2016, from: http://www.iau-hesd.net/sites/default/files/documents/copernicus.pdf.

Dominelli, L. (1997). *Anti-racist social work*. Basingstoke: Palgrave Macmillan.

Filho, W. L. (2000). Dealing with misconceptions on the concept of sustainability. *International Journal of Sustainability in Higher Education, 1*(1), 9–19.

Foucault, M. (1978). *The history of sexuality*. Baltimore: Penguin.

Hayward, C. R. (1998). *De-facing power*. Cambridge: Polity.

Hooks, B. (1994). *Outlaw culture: Resisting representations*. Abingdon: Routledge.

IAU (International Association of Universities). (1993). *Sustainable development and the universities: IAU's position*. Retrieved from November 20, 2016, from http://www.unesco.org/iau/sd/position.html.

Jacobs, M. (2016). High pressure for low emissions: How civil society created the Paris climate agreement. *Juncture, 22*(4), 314–323 (Chicago).

[16]https://www.theguardian.com/commentisfree/2016/dec/06/trump-world-climate-change-denial.

Ki-Moon, B. (2015). UN chief thanks civil society for its role in Paris Agreement. Retrieved November 26, 2016, from http://www.un.org/sustainabledevelopment/blog/2015/12/un-chief-offers-thanks-to-civil-society-for-its-role-in-paris-agreement/.

Latour, B. (1992). Where are the missing Masses? Sociology of a few mundane artefacts. In W. Bijker, & J. Law (Eds.) *Shaping technology, building society: Studies in sociotechnical change* (pp. 225–258). Cambridge: MIT Press.

Lozano, R., Lukman, R., Lozano, F. J., Huisingh, D., & Lambrechts, W. (2013). Declarations for sustainability in higher education: Becoming better leaders, through addressing the university system. *Journal of Cleaner Production, 48,* 10–19.

Maastricht University Green Office. (2016). Accessed December 8, 2016 http://greenofficemaastricht.nl.

Remnick, D. (2016). Obama Reckons with a Trump presidency. New Yorker. Retrieved December 08, 2016, from http://www.newyorker.com/magazine/2016/11/28/obama-reckons-with-a-trump-presidency.

RootAbility—making sustainability sustainable. (2016). Retrieved December 09, 2016, from http://rootability.com/.

Sukumaran, B., Chen, J., Mehta, Y., Mirchandani, D., & Hollar, K. (2004). A sustained effort for educating students about sustainable development. In *Proceedings of the 2004 American Society for Engineering Education Annual Conference and Exposition, Session* (Vol. 1793, pp. 1–8).

ULSF. (1990). *The talloires declaration*. University Leaders For A Sustainable Future, Washington, DC.

United Nations Educational, Scientific, and Cultural Organization. (2009). *UNESCO World Conference on Education for Sustainable Development—Bonn Declaration, Bonn.* www.esd-world-conference-2009.org/fileadmin/download/ESD2009_BonnDeclaration080409.pdf.

Vellani, F., & Nanjee, N. (2016). Institutionalizing ethics of sustainability in American Universities. In *Challenges in higher education for sustainability*. Berlin: Springer.

World Commission on Environment and Development (WCED). (1987). *Our common future—the brundtland report*. Oxford: Oxford University Press.

Wenner, J. (2016). *The day after: Obama on his legacy, Trump's win, and the way forward. Rolling Stone*. Retrieved December 08, 2016, from http://www.newyorker.com/magazine/2016/11/28/obama-reckons-with-a-trump-presidency.

Preschools Teachers' Sustainable Competencies Within New Kosovo Curriculum Framework

Arlinda Beka

Abstract Preschool Teacher Education and Sustainability has been a matter of discussions and researches among scholars and politicians worldwide for the last several years. Sustainability in Higher Education (including the Teacher Training) are seen by scholars and researchers as interlinked and interdependent with economic development and progress. Teacher Education in Kosovo following global trend in developing education has taken steps toward making this crucial link between Teacher Education and Sustainability in Higher Education. After several years of work in education reforms Kosovo authorities introduced the Kosovo's New Curriculum Framework in 2011. This has been a major step in orienting Kosovo's Education toward a competency based learning outcomes. The new Curriculum provides a set of guidelines for Teacher Competencies and the Teacher Education programs. It empowers the Faculty of Education of the University of Prishtina to provide the appropriate Teacher Training Programs according to the new Curriculum framework. This research paper will give a closer look to the Preschools Teachers' sustainable competencies within New Kosovo Curriculum Framework and their implementation. It is a first time that a research is done on this topic in our country and the outcomes will help the improvement of Teacher Education Programs of the Faculty of Education. This research paper was conducted using policy analysis of documents, relevant documents, KCF, Teaching Curricula at Faculty of Education, etc.

Keywords Teacher education · Competences · Sustainability · Curriculum

---

A. Beka (✉)
Faculty of Education, University of Prishtina, Prishtina, Kosovo
e-mail: arlinda.beka@uni-pr.edu; arlinda.beka@gmail.com

J.P. Davim (ed.), *Curricula for Sustainability in Higher Education*,
Management and Industrial Engineering, DOI 10.1007/978-3-319-56505-7_4

1 Introduction

There is a noticeable increase in interest with regards to sustainability in higher education. Following the global trends of development and increase of links between education and economic growth or employment countries and societies are paying much more attention to long term solution and sustainable future. Sustainability in higher education has been an issue that took a lot of attention among scholars, education policy decision makers and other stakeholders involved in Higher education or education in general. This becomes more relevant in the last two decades which are marked with the focusing in sustainability and ESD as an interdisciplinary approach. According to de Haan it is clear the interdependent relation between Education and Sustainable Development (de Haan 2006). Teaching for Sustainability is a process by which teachers turn the broad conceptual and often philosophical ideals of education for sustainability into specific elements of effective teaching practice (Nolet 2015).

Following this development in sustainability and education many countries have initiated or developed reforms in order to adjust with new changes and sustainability agenda. Reforms have been often structural and the politics of structural reform in most countries and in higher education shared the same basic characteristics. National politicians initiated change by attracting attention to problems with the education structure and building reform coalitions. Given chaotic and risky environments, elite bureaucrats allied themselves with politicians and supported reform in order to manage it. The bureaucrats networked in policy circles to shape the terms of debate and legitimize structural reforms (Nitta 2008).

So mainly the reforms start by politicians who often are not very much aware of the overall process, it impact to teachers, students and the entire generations. For them is more an political agenda (which may be a good one) that they want to push forward during the time they are in power, which often is not enough for completing a education reform process. Reforming education throughout society is never easy; major reforms usually are encouraged and accompanied by ideological, political and economic differences. For this reason, for the analysis of aspects of reform in a particular environment is important to look at a broader societal context beyond education (Cummings and Williams 2008). This is true since the education policy both reflects and reinforces national values, beliefs and purposes. Yet it is arguable that what unites developed countries is more important than the historical, cultural and social differences among them. In their thinking and approach to problems, tertiary education policy-makers often have more in common with their cross-border counterparts than with fellow nationals from outside their specialty (Taylor 2003). There is a difference among countries, and between K–12 and higher education, primarily occurred because national politicians varied in their attention to structural reform and, to a lesser extent, because interest groups varied in their cohesiveness. Countries in which national politicians paid strong, consistent attention to structural reform experienced significant restructuring, particularly in curriculum and admissions. In countries with weak or sporadic attention from national politicians, little structural education reform occurred, and what was enacted revolved around bureaucratically-initiated budgetary restructuring (Nitta 2008). Keith A. Nitta continues to explain that "structural reform also followed the same political pattern around the world. Faced with intractable problems such as budget deficits, politicians symbolically distanced themselves from the administrative system, blaming it for the problems of the day. They then drew on trendy managerial reforms to address the perceived structural problems. The similar politics of structural reform produced familiar loose–tight reform strategies (Nitta 2008). As we are talking about higher education (Teacher Training) we should bear in mind that in order to be sustainable, higher education today must be treated as a business. To be resilient to the problems it faces, higher education today should also be regarded as a marketable service with qualified, knowledgeable graduates who will contribute positively to society (Mathaisel and Comm 2013). An important part of reforms and changes in education system throughout the world is developing new curriculum to incorporate the changes education reform and policies bring in. The curriculum is the core of any substantial change in education, a vehicle through which the changes are being implemented. The science of curriculum has evolved from concerning a program of study to the content of that program, and more recently to the ways the content is taught (Egan 2003).

A curriculum can impact on a student teacher through its openness, emphasising what might becalled an ethics of remembering that would involve thinking beyond the sole and the coherent. Teachers may then not assume control of language and education, and instead embrace the poeti, including the poetry of scientific and economic thoughts. Scientific, economic and technical viewas on education are not as bad or as incongruent to early childhood education as has beenassumed by Rousseau, Froebel, Steiner, and countless critics of the governing of early childhood education in the late twentieth and early twenty-first centuries (Gibbons 2011). Education for sustainability aims to help learners develop new ways of thinking collaborating and solving problems so that can effectively engage with challenges (Nolet 2015).

So the challenge for both teacher education and government is to understand the meaning, application, and impact of coherence. How do teacher educators know that coherence exists in relation to a course they provide (from design to assessment and evaluation), and what might thatyou expect as an outcome of such coherence? A problem for government is the determining of anational horizon for teacher education that is suitably broad in its capacity to cope with thehorizons that are closer to the lived experiences of early childhood education and teachereducation (Gibbons 2011).

Modern days Curriculum, have a "competency based approach". They are to developed certain competences are predefined in Curriculum frameworks or other relevant documents in one country. Competencies are the positive combination of knowledge, ability and willingness in the availability of the individual to cope successfully and responsibly with changing situations (Weinert 2001). The term "competency" echoes throughout the country. It is discussed not only within the work environment or in the context of educational issues, but has also become a concern in personal and societal everyday life. Societal change, the progress of technology and globalization are accompanied by new challenges which have to be mastered: increasing individualization and growing societal diversity, accompanied in parallel by expanding economic and cultural uniformity, the availability of a rapidly growing amount of information, as well as the necessity to cope with increasing complexity and uncertainties (Ruchen 2001). In using the general approach, competencies may be characterized as dispositions to self-organization, comprising different psycho-social components, existing in a context-overlapping manner, and realizing themselves context-specifically. They may be acquired gradually in different stages, and they are reflected in successful actions (Barth et al. 2007). According to Rychen and Salganik, the term "key competency" seems of importance as it represents a qualitative extension that points out the special significance of certain competencies. Key competencies are relevant across different spheres of life and for all individuals (Rychen and Salganik 2003). They do not replace domain-specific competencies which are necessary for successful action in certain situations and contexts. They rather bear a different, a wider focus, pooling different competency classes and being situated transversely to them. They comprise different domain-specific competencies and point out the most relevant competency fields (Rychen and Salganik 2003). Again de Haan in his famous article from 2006 gives a much clearer description of key competences. According de Haan, Gestaltungskompetenz" comprises the following eight key competencies:

(1) Competency in foresighted thinking;
(2) Competency in interdisciplinary work;
(3) Competency in cosmopolitan perception, trans cultural understanding and co-operation;
(4) Participatory skills;
(5) Competency in planning and implementation;
(6) Capacity for empathy, compassion and solidarity;
(7) Competency in self-motivation and in motivating others; and
(8) Competency in distanced reflection on individual and cultural models (de Haan 2006).

De Haan elaborates it more: "Gestaltungskompetenz means the specific capacity to act and solve problems. Those who possess this competence can help, through active participation, to modify and shape the future of society, and to guide its social, economic, technological and ecological changes along the lines of sustainable development. Gestaltungskompetenz, (…), means: having the skills, competencies and knowledge to change economic, ecological and social behavior without these changes merely being a reaction to existing problems. Gestaltungskompetenz makes an open future possible that can be actively shaped and in which various options exist". (de Haan 2006). So he really speaks of shaping skills which every society would like to be able to develop with their students. This kind of shaping skills can't develop with a few workshops or retreats, or even training sessions for teachers. In universities especially, teachers have something of an aversion to being trained, and above all on issues they do not necessarily see as being relevant to their own discipline. Nor do they want to be instructed 'top-down' on methodologies and the subject matter of their own lectures. If we really want to be effective in integrating elements of sustainable development in overall higher education, it is necessary for the lecturer to see the specific intellectual challenges that sustainable development poses to his or her discipline (Corcoran and Wals 2004).

Developing the ability of citizens to create sustainable, resilient communities—locally and globally—define the very purpose of schooling. The following fundamental understandings also drive this document:

- The way to increased prosperity for all citizens is through collectively addressing the critical environmental, economic, and social issues of our local communities.
- A—effective‖ teacher is one who helps all students achieve high levels of academic success, develops their personal and social wellbeing, and inspires meaningful involvement in their local social, economic and ecological communities.
- An effective teacher educator is one who prepares teachers to do the work outlined above.
- In order to be integrated into curricula, education for sustainability approaches must be implemented in ways that meet standards; this is creative work for teachers and teacher educators (Santone et al. 2004).

Kosovo itself has worked in bringing sustainability and ESD principles into national agenda of policy makers, academics and faculty scholars, and other stakeholders. "Kosovo's institutions have worked hard to improve their legislation in accordance with the contemporary trends of the western hemisphere, particularly base upon European Union (EU) guidelines and regulations, as part of their commitment to be part of the EU in the future. In this manner they have made changes in the education system aiming to improve the quality of education and reform the whole education system (Beka 2015). Kosovo has made major changes and improvements to its teacher training education including in-service and pre-service teacher training. Most of those change followed the changes in education system, education strategies and the Curriculum Framework, etc. Kosovo policy makers have decided to take education system through a change only a decade after the Curriculum was adopted at 2001. Even when the decision to make the curriculum reform in 2001 was taken it didn't count on all challenges and difficulties. "Development of the new Curriculum Framework of Kosovo in 2001 through which Kosovo education system aimed to comply with European and international course in education was not consistent with the circumstances that had passed Kosovo, as well as an emergency situation in which Kosovo was in those years. European space of education had consistently planned and applied reforms, while our country was in the survival stage of education. This is a big difference and completely unmatched, to follow the example of reform. Education reforms in any developed country weren't facing the same difficulties that were in Kosovo (Beka 2014a).

In Kosovo the Curriculum of 2001 lasted only a decade since the new Curriculum Framework was introduced at 2011 bringing major changes. It is important to note here while new KCF was introduced to Kosovo in 2011 all the training for teacher conducting until then were done based upon the Curriculum of 2001 focusing on how to implement it. This made the situation more difficult for the entire pore university education in Kosovo (Beka 2014b).

The current structure of educational system is 5 + 4 + 4 out of which are 9 years of compulsory education, Grades 1–5 primary education (class teaching), and 4 years (Grades 5–9) of lower secondary education (subject teaching). The preschool level will become obligatory for all children starting from 2015. The main educational document upon which this system relies is the Kosovo Curriculum Framework (KCF), which is a modern competence based curriculum MEST 2011). New KCF is focused in six areas of learning. The areas of learning are Language and Communication, Arts, Mathematics, Natural Sciences, Society and environment, Health and Welfare, Life and Work, which apply from preschool level to pre university level regardless of general education schools or professional education schools. The learning areas enable the learners to develop further knowledge and skills about themselves, others, and society in a wider sense (Beka 2015).

One of the main aims of education in Kosovo is the development of knowledge, skills, attitudes and values required by a democratic society. This will enable young people to become active and responsible citizens so that they deal constructively with the challenges of diversity, as well as cultivating and respecting their own rights and the rights of others (MEST 2011)

According to the Kosovo Curriculum Framework, the education system in Kosovo enables individuals to become independent, able to fulfill their personal life and to contribute to the continuous progress, prosperity and welfare of Kosovar society.

The aims of education are:

- The development of personal and national identity, statehood and cultural belonging;
- The promotion of general cultural and civic values;
- The development of responsibility for themselves, for others, for society and for the environment;
- Preparation for life and work in the context of social and cultural changes; development of entrepreneurship and use of technological skills;
- Preparation for lifelong learning (MEST 2011).

According to this relevant document, Kosovo Education should strive toward becoming a "knowledge society". Therefore the Kosovo Curriculum Framework has envisaged those key competencies that derive from from the general aims of pre-university education in Kosovo and define the main learning outcomes that learners need to achieve in a progressive and consistent way throughout the pre-university educational system. In compliance with the aims of education of Kosovo, the key competencies envisaged for the pre-university education system in Kosovo are:

Competency in communication and expression;
Competency in thinking;
Competency in learning;
Competency in life, work and environment-related areas;
Personal competency;
Civic competency." (MEST 2011)

Teachers are the one who should be able to transmit and develop those competences to their pupils. The KCF emphasizes the importance in developing those competences from the preschool education as starting point. "The aim of the educational process in this phase is to stimulate their creativity and enthusiasm in approaching new experiences and facing real-life situations. During this period, children will be encouraged to develop communication skills in their mother-tongue in simple everyday situations, to increase their attention and concentration skills, and to develop their basic social skills. Particular attention during this level will be paid to their physical development through physical activities and experiences that will increase the children's awareness of their physical well-being and their health and safety. In the pre-primary grade (ages 5–6), children will be exposed to learning experiences that include the basic elements of reading, writing and numeracy, thus raising their level of preparedness for subsequent schooling and life" (MEST 2011). In order to develop the education system in general the focus needs to be at the preschool education. From this level of education it can be achieved to offer to the children education and development of skills as a preparation for competence development. So in order for Kosovo children to gain the described competences based upon the KCF we are to focus in preschool education.

In order to achieve those competences, according to the KCF, for children (pupils) emphasize should be given to the preschool educators. Based upon demands of KCF, Ministry of Education has work in developing the professional competences of preschool educators through which they will be professionally competent to developed competences to the children.

Kosovo has adopted the competences for educators which have been developed by International Step by Step Association-ISSA. According to ISSA, the Competent Educators of the 21st Century consists of seven focus areas:

1. Interactions
2. Family and Community
3. Inclusion, Diversity and Values of Democracy
4. Assessment and Planning
5. Teaching Strategies
6. Learning Environment
7. Professional Development (ISSA 2010)

Faculty of Education following the changes of KCF has taken reorganize its study programs to adjust with those changes and prepare pre-service and in-service teachers in developing teaching competences based upon the requirements of KCF. Preparation of professionally competent educators is linked directly to the study programs that are offered during the four years of studies at Faculty of Education of University of Prishtina.

Within the Faculty of Education there are 2 programs that prepared educators to work with children of 0–6 years old.

First program is the "Early childhood education" 0–3 years and the second one is "Preschool education" 3–6 years. This division is made in order to offer children special focus according to their development stating for the age 0. At early childhood, 0–3 years children don't need just "baby care", as that has been proven it's not enough. Child development at this age is more meaningful and educators should work closely to enable that the child gets the best of physical, social, emotional, cognitive development that will help them in the future development.

Until recently with children of 0–3 age worked medical nurses who not necessarily had any preschool education. This has changed now and in order to be able to work at preschool with this age of children, it is require being a graduate from Faculty of Education, early childhood education program. Therefore the role of Faculty of Education in enabling the competent educators in crucial as well as the responsibility is big. Based upon Faculty of Education Strategic Development Plan, the mission of the Faculty is: "To provide quality study programmes for initial and continuous training of teachers and other specialists in the area of education, as well as to develop scientific-research activity, for the purpose of improvement of the Education System in Kosovo" (FEdu 2013).

Furthermore the Strategic Development Plan of the Faculty of Education has identified five strategic objectives set by the Planning Group, which are:

1. To develop and provide programmes in line with needs of the Education System,
2. To provide quality and effective services for students and programme attendees,
3. To improve scientific research activity in the Faculty,
4. To build capacity for quality teaching and scientific-research activity in the area of Education,
5. To build functioning and sustainable partnerships with local and international institutions (FEdu 2013).

Following the implementation of this Strategic Development Plan, Faculty of Education has developed study programs for early childhood education 0–3 and preschool education 3–6. Creation of study programs of the Faculty of Education has followed the required competences for educators which are divided in three categories:

1. Theoretical courses- through which students get different information for child development.

2. Courses that help development of skills for students (future teachers), through which students will be able to analyze and act upon situation or circumstances that they face in their work with children.

3. Professional practice, which is very important as it helps students to strengthen and implement their gained knowledge into practice.

However it still early to make the concrete measures of those study programs of Faculty of Education are the right ones in preparing of the competent educators.

2 Conclusion

Sustainability in higher education, in teacher training in our case, can be measured in different ways, by learning outcomes, employability of graduates, professional competences of future teachers, etc. Yet we know we are living in a changing world where new skills are required particularly for those who are working in educating new generations.

In Kosovo we have now two standards to measure and compare the development of educators competences. While the KCF clearly emphasizes what competences are expected for children according to their age and their development. Educators' competences in the other hand shows what are the competences that educators are to develop in order to get the best for children they work with.

Faculty of Education is working in implementing both documents by preparing the future educators to be up to the task, competent and help children develop their competences as the grow in maturity and knowledge.

Recommendations
To MEST:

- To continuously oversee the implementation of KCF and intervene immediately if needed.
- To help schools, preschool in providing assistance and support in implementing the KCF.
- To support Faculty of Education in further professional development of it staff as important factor in the teacher training process.
- To help Faculty of Education in conducting researches and measures of its study programs

To Faculty of Education:

- To strengthen its study programs with up to date researches
- To intervene whenever is needed in adjusting study programs, curricula or syllabi in order to prepare and develop competent educators and teachers
- To cooperate closely with preschools and schools so it has information from the ground on implementation of their study programs.

To Educators:

- To consider their continuous professional development as a very important part of the career development.
- To expand their knowledge and competences regularly as part of their professional growth.
- To understand that competence is a process and not a point.

References

Barth, M. et al. (2007). Developing key competencies for sustainable development in higher education. *International Journal of Sustainability in Higher Education, 8*(4), 416–430.

Beka, A. (2014a). Reforming the Education System in Kosovo with the Start of the 9th Classes. *Journal of Educational and Social Research, 4*(1).

Beka, A. (2014b). Partial or comprehensive change of system- challenges of education reform in Kosovo. Unpublished Ph.D disertation, Prishtina.

Beka, A. (2015). The Kosovo education for sustainable development's role in promoting the decade of education for sustainable development in Kosovo. *Applied Environmental Education & Communication*, 126–131.

Beka, A., & Ciani, A. (2015). Promoting Education for Sustainable Development at University of Prishtina. In W. Leal Filho et al. (eds.), *Integrating Sustainability Thinking in Science and Engineering Curricula* (pp. 139–151). Cham: Springer International Publishing. DOI 10.1007/978-3-319-09474-8_11.

Corcoran, P. B., & Wals, A. E. J. (Eds.). (2004). *Higher education and the challenge of sustainability: Problematics*. Dordrecht: Kluwer Academic Publishers.

Cummings, W. K., & Williams, J. H. (2008). *Policy-making for education reform in developing countries*. UK: Rowman & Littlefield Education.

de Haan, G. (2006). The BLK '21' programme in Germany: a 'Gestaltungskompetenz'-based model for Education for Sustainable Development. *Environmental Education Research, 12*(1), 19–32.

Egan, K. (2003). What is curriculum? *Journal of the Canadian Association for Curriculum Studies, 1*(1), 9–16.

FEdu. (2013). Faculty of Education Strategic Development Plan. Prishtina: Faculty of Education.

Gibbons, A. N. (2011). The incoherence of curriculum: Questions concerning early childhood teacher educators. *Australasian Journal of Early Childhood 36*(1).

ISSA. (2010). http://issa.nl/content/issa-quality-principles. Retrieved 2016, from http://issa.nl/content/issa-quality-principles: http://issa.nl.

Mathaisel, D., & Comm, C. (2013). A syllabus for resilience in higher. In S. e. (eds), *Sustainability Assessment Tools in Higher Education Institutions* (pp. 291–307).Cham: Springer.

MEST. (2011). http://www.ibe.unesco.org/curricula/kosovo/kv_alfw_2011_eng.pdf. Retrieved 2016, from http://www.ibe.unesco.org/curricula/kosovo/kv_alfw_2011_eng.pdf: http://www.ibe.unesco.org/curricula/kosovo/kv_alfw_2011_eng.pdf.

MEST. (2016). http://masht.rks-gov.net/uploads/2016/02/raport-vleresimi-psak-2011-2016-eng-web.pdf. Retrieved 2016, from http://masht.rks-gov.net/uploads/2016/02/raport-vleresimi-psak-2011-2016-eng-web.pdf: http://masht.rks-gov.net/uploads/2016/02/raport-vleresimi-psak-2011-2016-eng-web.pdf.

Nitta, K. A. (2008). *The politics of structural education reform*. New York: Routledge.

Nolet, V. (2015). *Educating for sustainability*. New York: Routledge.

Ruchen, D. (2001). Defining and selecting key competences. In D. &. Rychen, *Introduction* (pp. 1–16). Seattle: Hogrefe & Huber.

Rychen, D. S., & Salganik, L. H. (Eds.). (2003). *Key competencies for a successful life and a well-functioning society*. Göttingen: Hogrefe & Huber.

Santone, S., Saunders, S., & Seguin, C. (2004). Essential Elements of Sustainability in Teacher Education. *Journal of Sustainability Education.*

Taylor, W. (2003). Steering change in tertiary education. In e. Heather Eggins, *Globalization and Reform in Higher Education* (pp. 11–32). England: Open University Press.

Weinert, F. (2001). Concept of competence: A conceptual clarification. In D. R. (eds), *Defining and Selecting Key Competences* (pp. 45–66). Seatle: Hogrefe & Huber.

Building Energy, Environment and Sustainability Linkages in Management Education in India—An Innovative Curriculum Based Approach

Prakash Rao, Yogesh Patil, Viraja Bhat and Manisha Ketkar

Abstract In the past few years the focuses on linkages between global energy, environment and developmental issues have caught the attention of various sections of business and industry. The Earth Summit of 2012 has further strengthened the case for integrating sustainable development as an integral part of higher education institutions through a process of cohesive learning and use of innovative pedagogical methods including the internationalization of educational systems. The present paper seeks to focus on designing and implementing a unique industry relevant curricula on energy and environment integrating various aspects of sustainable development by building institutional capacities in key sectors like energy, environment, infrastructure, agriculture, climate change, global governance and law and corporate sustainability at the Symbiosis International University. The emphasis is on evolving an approach for embedding sustainability education in International Business and related areas by matching the needs of business and industry. The paper through a case study at a constituent institute of Symbiosis International University explores the linkages between energy, environment, development and corporate sustainability issues and suggests an innovative approach for business schools to adopt an integrated curriculum. Key areas that have been suggested in the integration process in the context of a growth cantered economy include power, water–energy nexus, industrial ecology, etc. The paper also discusses sustainability education using interdisciplinary thematic areas like

P. Rao (✉) · Y. Patil · V. Bhat
Department of Energy and Environment, Symbiosis Institute of International Business, Symbiosis International University, Pune, India
e-mail: dydirector@siib.ac.in

Y. Patil
Symbiosis Centre for Research and Innovation, Symbiosis International University, Pune, India

M. Ketkar
Symbiosis School of Banking and Finance, Symbiosis International University, Pune, India

J.P. Davim (ed.), *Curricula for Sustainability in Higher Education*, Management and Industrial Engineering, DOI 10.1007/978-3-319-56505-7_5

international trade and energy policies, environmental law, environmental economics, global and private voluntary standards, climate mitigation, etc. In conclusion the paper outlines elements for developing an interdisciplinary approach for integrating sustainable development in management education particularly in emerging markets.

Keywords Business school · Curriculum · Energy · Environment · Management education · Sustainability linkages

1 Education for Sustainable Development

Education has played a key part in driving the paradigm of growth across post independence India. Since 1947, Independent India has had a major thrust towards industrialisation and agriculture, and building the country's granaries to fulfil the needs of its citizens was a key priority. Building a team of qualified professionals across the rest of the country to serve the needs of the industrialized world was a key priority of the policy makers. India's English literate population is a key to the competitiveness (Porter 1990).

As a part of the Nation building activities several important initiatives were taken up aimed at increasing the knowledge base of the country through the establishment of specialised Institutions like the Indian Institutes of Management (IIMs) and the Indian Institutes of Technology (IITs). Simultaneously, other educational institutions also started developing their own centres aimed at imparting quality training and knowledge based education. Today, these numbers have grown exponentially. In the higher education sector, India today has a total of over 750 universities, 43 central universities, 299 state universities, 140 private Universities, 128 deemed universities and 5 institutions established through state legislation, 30 Institutions of national importance. The total government spending on education is approximately 3.8% of its GDP on education and FDI inflows in the education sector during May 2012 stood at $31.22 million.

The concept of sustainable development gained worldwide support through the publication of the Brundtland report "Our common Future" by the World Commission on Environment and Development in 1987. Subsequently, the 1992 earth summit further saw the emergence of education as major thrust areas in fostering environmental sustainability through the aegis of Agenda 21, which was a charter of principles aimed at promoting education, public awareness and training. The first decade of 2000 saw the world summit on sustainable development at Johannesburg in 2002, which recommended amongst other issues, the internationalisation of educational systems at all levels of learning. Consequently, the UN declared the decade of 2005–2014 as the "Decade of education for sustainable development". The need for a new paradigm of education in sustainable development, given the rapid pace at which humanity was using up the planet's resources required an education framework to address critical sustainability and environmental issues at Higher Education Institutions (HEI).

Sustainability itself, however, is a complex term that has been open to a variety of interpretations (Bonnett 2002). In the past, although the concept of sustainability tended to be centred squarely on environmental issues, as in the framing of a balance between economic growth and ecological carrying capacity, more recent understanding of the term have come to include socio-cultural aspects as well (Dempsey et al. 2009). Approaches to sustainability-related issues are diverse and situations are uniquely different between nations and cultures (Ehrenfield 1997; Matten and Moon 2004).

In recent times, apart from incorporating sustainability related curriculum in educational institutions, it has also been suggested that the institutions of higher learning should also measure their sustainability quotient in terms of tracking various energy and ecological foot prints and impacts (Leal Filho 2012). A sustainable business management practice has to acknowledge the "embeddedness in social, environmental and economic systems, and focuses on management and relationships to meet the environmental, social, and economic requirements of many different stakeholders in its networks" (van Kleef and Roome 2007).

Academic institutions have in recent years apart from focusing on sustainable development based curriculum are also considering the prospect of incorporating some of the emerging disciplines into their own sustainability models. Sterling (2001) points out that development of sustainable education framework requires indepth visioning, design, and implementation at levels for achieving environmentally sustainable lifestyles. Several other authors (Madeira et al. 2011; Mitchell 2011; Speller 1992; Stubbs and Cocklin 2008) have also highlighted the need for incorporating sustainable development into the framework of University institutions. Developing key competencies in sustainability is necessary to shape new and emerging academic programs where real world issues and challenges are often discussed. Recent studies have also suggested a convergence based approach which enables students to analyse sustainability problems as well as to anticipate future sustainability challenges (Wiek et al. 2011).

The desire to achieve sustainable development often has its fair share of challenges. Often educational institutions are seen to have certain misconceptions of sustainability and its implications for society. While most would agree that environment and conservation activities are seen as a positive sign of societal development, serious challenges remain in implementation of such actions. Some of the barriers to such actions include resource availability, lack of relevant human resource capacity, motivation among institutional staff about environment issues, management support, etc. (Leal Filho 2000, 2011).

In India, a few business schools and academics institutions have begun to introduce and also altering existing curriculum to meet not only the requirements of students but also match the needs of business and industry for sustainability initiatives (Park et al. 2012; Tikoo 2009). The inclusion of integrating environment as an important component of corporate sustainability efforts has only been a recent development across Indian business schools with limited intervention.

2 Industry Issues and Approaches in Sustainable Development

With the adoption of the Rio +5 declaration in 2002, various approaches were developed globally by academic institutions in designing and operationalizing sustainable development activities at Universities through either legislations, regulatory frameworks, policies or voluntary standards. The Washington based Aspen Institute has created a systematic ranking system of Business Schools around the world which have taken on sustainability initiatives (Godemann et al. 2011). The effort to align Industry based standards like Global Reporting Initiative (GRI) with sustainability practices at Universities is an important step towards sustainable development as a core activity at Universities (Lozano 2011). It is also true that while more than 600 Universities worldwide have committed themselves to various aspects of promoting sustainable development, very few have been able to successfully integrate the principles of sustainable development into curriculum and practice due to a combination of reasons, varying from lack of awareness, low level of staff involvement, lack of capacity building amongst faculty (Leal Filho 2011; Lozano and Watson 2013). In the Asia Pacific regional attempts have been made to get a preliminary understanding of trends in sustainability in business education (Naeem and Neal 2012) (Table 1).

The emergence of major environmental issues across the world has led to a series of multi-lateral treaties for addressing these issues. Simultaneously, educational institutions around the world also realised the importance of incorporating environment and sustainable development issues as part of curriculum development and transformation. Post the Rio Summit of 2002, and the Paris climate negotiations in 2015, there has been a renewed interest in developing capacities of professionals in diverse environmental issues like natural resource management, energy conservation, climate change, renewable energy, environmental economics, ethics, and corporate sustainability etc. However, while governments, business and industry, civil society organizations, technologists, research and development organisations, developmental sector organisations and activists were at the fore front of addressing these issues for tackling critical issues of the environment, the academic community had an onerous task of building suitable and relevant curriculum and course content to provide skilled human resources for diverse industry sectors. The role of management institutions and business schools were therefore seen as a vital part of nation building process particularly for building managerial capabilities in sustainability across different dimensions.

The importance of Environment, Health and Safety (EHS) standards and monitoring has in recent times gained prominence as a key aspect of sustainability for industries. Increasing the pitch of globalisation in emerging markets and the need to adopt cleaner technologies has further given an impetus for business and industry to be more proactive and move into a system of high integration rather than of compliance. Informal market analysis seems to suggest that there is considerable

Table 1 Management schools and industry oriented institutes offering Energy and Environment based curriculum and courses in India

Sr. No.	Institution[a]	Course	Duration	Degree
1	Indian Institute of Management, Ahmedabad	Carbon trading, Carbon finance	One year	Diploma
2.	Amity School of Natural resources and Sustainable development, Noida, Delhi	Natural resource management	Two years	MBA
3.	DeenDayal Petroleum University, Gandhi Nagar, Gujarat	Oil and gas, Petroleum management, Petroleum Economics	Two years	MBA
4.	Symbiosis Institute of International Business, Symbiosis International University, Pune	Energy and environment	Two years	MBA
5.	Indian Institute of Forest Management, Bhopal	Forestry management	Two years	Post Graduate Diploma
6.	Management Development Institute, Gurgaon	Energy management	Fifteen months	Executive Post Graduate Diploma
7.	National Institute of Industrial Engineering, Mumbai	Industrial safety, Environmental management	Two years	Post Graduate Diploma
8.	Indian Institute of Social Welfare and Business Management,Kolkata	Public Systems management	Two years	Masters degree
9.	TERI University, New Delhi	Business sustainability	Two years	MBA
10.	University of Petroleum and Energy Studies, Dehradun	Oil and gas management, Energy trading	Two years	MBA
11.	Rajiv Gandhi Institute of Petroleum Technology, Rae Bareli	Petroleum and Energy management	Two years	MBA
12.	Management Development Institute, Gurgaon, Haryana	Energy management	Fifteen months	PGP in Management
13.	National Power Training Institute, Faridabad, Haryana	Power management	Two years	MBA
14.	Institute of Energy Management and Research, GLIMS, Haryana	Energy management	Two years	PGP in Management
15.	Chh. Shahu Institute of Business Education and Research, Kolhapur, Maharashtra	Environmental management	Two years	MBA

[a]Only an indicative list; Rao et al. (2013)

potential for capacity building for managers and factory workers in this area in the Asian region.

At the same time Industry based associations like Confederation of Indian Industry (CII), FICCI, and World Business Council for Sustainable Development (WBCSD) have strongly advocated sustainability as core to their operational practise but also gone on to promote capacity building in the value chain amongst suppliers and vendors.

Market driven approaches, favourable government polices and global sustainability mandates have been the trigger for businesses to rethink their long term strategies towards building local and regional capacity for their workforce. As a result, emerging domain areas like renewable energy, energy efficiency, sustainable transport, green building architecture etc. have found favour with many industries as possible sunrise sectors. The approach of a compliance based action by industry and lack of environmental awareness amongst could therefore be an impediment for achieving sustainable growth. This is also due to the fact that Industry has often perceived environmental issues as a regulatory mechanism rather than as a business case. The argument seems to have some credence as investing in natural capital is seen as a cost centre and therefore not significant for industry growth. In a sense, policy decisions favouring environmental conservation in the past have been presumably seen to undermine Industry confidence for expansion and tapping new and emerging markets.

The present initiative in developing an integrated curriculum combines elements of technological advancement, social equity, economic viability, regulatory frameworks and global strategies as the basis for building a holistic sustainability education paradigm (Rao et al. 2013).

3 The SIIB Integrated Energy and Environment Programme

In 2009, the Symbiosis International University, Pune through its constituent institute, the Symbiosis Institute of International Business, developed a post graduate programme which was aimed at integrating environmental sustainability, energy production, and economic growth in curriculum. The programme provides for a holistic understanding of developing competencies in emerging technologies, economic issues and global environmental strategies. Core focus areas include sectors like sustainable energy development, renewable energy, power economics, climate change, carbon financing and markets, corporate sustainability, environmental impact assessments, natural resources management, etc.

The University in 2013 also embarked on a curriculum development exercise aimed at standardisation its curricula across various faculty disciplines including the sustainability and infrastructure area (Table 2). The effort was meant to integrate

Table 2 Common courses in Sustainability and Infrastructure at Symbiosis International University

Sr. No.	Common courses on Sustainability across Symbiosis International University	UG/PG
1.	Concepts and Applications in Sustainability	PG
2.	Environment Impact Assessment	UG/PG
3.	Oil and Gas Economies	PG
4.	Sustainability standards—Application, Analysis and Reporting	PG
5.	Public Private Partnerships	PG
6.	Field/Research Project	UG/PG
7.	Governance and Corporate Sustainability	PG
8.	Project Management Suite	PG
9.	Transmission and Distribution Management	PG
10.	Urban and Industrial Waste Management	PG
11.	Renewable Energy Sources	PG
12.	Climate change, Carbon Markets and Financing	PG
13.	Environment Management Systems	PG

several courses which otherwise were seen as stand-alone courses with no proper linkages across the constituent institutes of the University.

While most educational institutions have developed niche based courses with specific focus on creating particular domain knowledge for students, the present programme was conceived with a view to integrate some of the critical issues linking the environment sector with energy development as well as current issues of social development and equity. Some of these integrated approaches are detailed below:

The programme developed in 2009, is a full-fledged post graduate course around what we term the "centrality" approach in designing and conducting industry oriented courses. This approach focuses on using key industry relevant issues around which a cluster of courses are developed which could then provide for a holistic view to integrate not only the technology aspects and issues but also focuses on a business and societal angle. We found several such issues and initiatives which formed the core of the programme around which other minor courses could be integrated. A few of these issues or programme areas which are critical aspects of sustainability of business and industry are detailed below:

3.1 Electricity Governance, Economics and Trading

Recent changes in power sector in the country following the Electricity Act, 2003 has mandated the development of a transparent, fair and equitable process in the generation, transmission and distribution of electricity across the country. This has led to several private sector entities trying to capture a market share of the power sector business. The post graduate programme has tried to include these issues as part of a holistic curriculum which brings in course elements using electricity as a central focus around which technology, regulatory issues, consumer needs, social equity have been factored in. At the same time related impacts on environment from power infrastructure have also been considered in the curriculum e.g. environmental impact assessments rules and regulations, wastewater management, etc. This is attributed to the fact that environmental concerns like mining, land acquisition, loss of ecologically rich ecosystems, wetland reclamation, forest cover loss etc. have been often neglected in the desire to set up new and ultra-modern power plants (Areendran et al. 2013).

3.2 *Corporate Sustainability*

The increasing rate of urbanization in the Asia Pacific geographies and local environmental stresses from over population, industrial development, migration etc. (Mukhopadhyay and Revi 2009) are only a strong indicator of how important environmental concerns are for businesses to adopt. Building a strategic involvement of regulatory institutions, business and industry, research institutions and civil society in an alliance will help create an economy which will foster a low carbon growth economy (Saqib et al. 2007).

The current initiatives in Indian industry to build environmental sustainability as a key mandate of businesses is limited and seem grossly inadequate given the enormous efforts needed to tackle global and domestic problems in the environmental, social, economic and technological space. The adoption of the triple bottom concept of environmental, social and economic performance of a business entity towards total sustainability can go a long way in ensuring both environmental and economic security amongst corporate sector. Education content with regard to corporate social responsibility and governance is now seen to be a major aspect of most business schools (Matten and Moon 2004).The corporate sustainability curriculum at SIIB incorporates the business case of sustainability through industry best practices. The curriculum goes on to include global sustainability standards as the hub around which elements of governance, social inclusion, environmental performance are closely woven. This is expected to help professionals to develop clarity and thinking on the balance between infrastructure growth and sustainable development through use of methodologies and tools for developing sustainability standards and guidelines.

It is widely recognised that business and management schools across the world accord low priority to the social responsibility while decision making as the approach is mainly towards narrow shareholder value ideology. However, in recent times, the discourse on Corporate Social Responsibility (CSR), ethics and governance has led to the inclusion of several such courses across many universities and autonomous higher education institutions. Education content with regard to corporate social responsibility and governance is now seen to be a major aspect of most business schools (Matten and Moon 2004).In an emerging economy like India, there is a greater need to develop awareness amongst students of societal responsibilities and build the capacity of young professionals on these issues. The business schools therefore need to engage themselves in real world development of society apart from classroom teaching. In line with the vision of Symbiosis and for the holistic development of students, the Symbiosis Programme conducts societal impact courses like CSR not only as a curriculum centred approach but also through practical and real time field understanding. The corporate sustainability curriculum incorporates the business case of sustainability through industry best practices. The curriculum goes on to include global sustainability standards as the hub around which elements of governance, social inclusion, environmental performance are closely woven (Christensen et al. 2007).This is taken up through the project approach and is monitored year on year on the path to achieve sustainable development. Key areas of engagement include education and environment (monitoring onsite energy, water, waste management measures activities, biodiversity, resources use, etc.). In the education area, focused class room based CSR courses are handled through practical field experience when students are assigned tasks to develop capacities of young school children of under privileged families. The social impact assessments of these initiatives are being monitored to demonstrate the effectiveness of these courses.

3.3 *Water-Energy Nexus*

In the road to sustainable development, water and energy related services and management are rated very high not only amongst businesses but also among scientists, policy makers, environmental professionals and the average citizen. High water consumption patterns in key business sectors as well as in urban areas coupled with inefficient energy technologies have often led to serious environmental concerns (Rao and Patil 2016). While most knowledge based sectors discuss water and energy management as separate entities, the curriculum at Symbiosis has attempted to integrate the two disciplines in an effort to provide a holistic approach to the water energy nexus (Rao et al. 2013). The courses have been created by studying the water-energy use through demand side management options, optimizing efficiency of existing systems e.g. water savings through harvesting and withdrawal, recycling of waste water and alternate energy use.

3.4 *Climate Change and Energy Development*

In recent times the issue of Climate Change is seen as perhaps the most important global environmental threat affecting fabric of society and economic growth. This is likely to have tremendous impacts on biodiversity, natural resources, freshwater resources, local livelihoods and many other sectors. From a business perspective serious challenges are seen as many industries rely on natural resources for production and operations. The Symbiosis programme has tried to integrate a low carbon development based curriculum (Date-Huxtable et al. 2013) within courses in the management discipline through the direct linkages between global climate change impacts and energy development but also as an enabler for students to demonstrate low carbon activities through practice. The pedagogical approach uses assessments of sectoral Greenhouse Gas (GHG) emissions through class projects and real time exercises using the Institute emissions as a live example. The results and analysis are then interpreted at national and global institutional levels. This enables students to integrate local-national-global climate related issues with GHG-Energy use-low carbon economy. The curriculum using the centrality approach also delves into UN policy processes of mitigation and adaptation following the post Paris (COP21) climate negotiations including an understanding of the Intended Nationally Determined Contributions (INDCs) for signatories to the UNFCCC.

The programme at SIIB also uses the approach to include courses tied to market driven scenarios and brings in recent industry and regulatory driven approaches in curriculum development. This hinges on innovative market based mechanisms which might make industry sectors lucrative in terms of return on investment apart from achieving the objective of sustainable development. Such mechanisms offer vast scope for various business and industry players including small and medium enterprises (SMEs) (Saini et al. 2012). Courses related to the recently established PAT scheme (Perform, Achieve and Trade) of the Government of India through the Bureau of Energy Efficiency is one such step which is likely to provide a major impetus to strengthening energy efficiency practices. Similarly, a course on the concept of Renewable Energy Certificates (REC) is also a recent attempt to involve industry players at providing market based incentives to project developers in the renewable energy space in increasing the overall share of renewable energy in the overall energy mix of India.

3.5 Public Private Partnerships in the Energy-Environment Sector

The Symbiosis Integrated programme has also introduced an innovative course module which seeks to conceptualise the governance model of public private partnerships as a solution to rapid economic growth. While this has had mixed results globally, it is of interest to note that several PPP lead initiatives have been taken up in water and sanitation, sustainable transport, power sector to promote sustainable infrastructure growth. We believe that the PPP led governance model with a focus on sustainable development is of great relevance as a course element for management education.

3.6 Industrial Ecology

Since long time, economic development has been a subject of great concern and a mechanism to augment social prosperity. In the process of economic development, actions like safeguarding, protecting and conserving the environment were completely missed out (World Bank 2000). This resulted into massive environmental deterioration and resources depletion at the global level. In the era of climate change and other global issues, its impacts on ecology got further intensified (Patil and Rao 2015).

Considering this deficit, the present Energy and Environment integrated programme at Symbiosis decided to introduce a new integrated industrial planning and management mechanism in its curriculum and 'Industrial Ecology' was one such emerging concept in the evolution of environmental management paradigms (Ehrenfeld 1997) and springs from interests in integrating notions of sustainability into environmental and economic systems (Ehrenfeld 1997).

Frosch and Gallopoulos in 1989 first proposed the idea of Industrial Ecology (Heeres et al. 2004) and since then the overall industrial development world across has entered a new perspective of production and process system. Fundamentally, industrial ecology is a novel approach to the industrial design of products and processes and the implementation of sustainable manufacturing strategies. Under this concept, the industrial system is interlinked with the surrounding industrial systems and in no circumstances works under isolation. It seeks to optimize the total materials cycle from virgin material to finished material, to component, to product, to waste product, and to ultimate disposal (Jelinski et al. 1992).Thus, the industrial development should resemble the natural ecosystem because in such a system, energy and resources are optimally used and wastes are absent. In contrast to the conventional concept, industrial ecology considers industrial waste as an economic resource (Ehrenfeld and Gertler 1997), which upon reducing, reusing and recycling, recovering means greater profit to the industries/companies in that ecosystem. In other words, mimicking natural ecosystem is the key for designing Industrial Symbiosis (Garner and Keoleian 1995) because natural ecosystems are self-contained, self-sustained and generate zero waste through complex interactions of food chains. Waste materials and energy originated from one industry may be used as a feedstock by other industry. Therefore, without much or very little investment in technology for waste management in developing and underdeveloped countries and waste can become a source (Patil 2012). Industrial ecology will thus help the industries to become more competitive by way of improving their environmental performance and strategic planning. Secondly, it will help local communities to develop and maintain a sound industrial base and infrastructure without sacrificing the quality of their environments; and thirdly it will help local, regional and national government to formulate policies and regulations in order to improve environmental protection with simultaneous building of business competitiveness.

Extending the industrial ecology concept further, we have also pitched in Life Cycle Assessment (LCA) in the course content as a method to assess environmental impacts associated with all the stages of a product's life from cradle-to-grave (Finnveden et al. 2009). LCA avoids narrow outlook towards the environment by way of: (a) Accounting an inventory of relevant energy and material inputs and environmental releases; (b) Evaluates the potential impacts associated with identified inputs and releases; and (c) Interpreting the results to help make a more informed decision (Hubbard and Bowe 2010). LCA is known to be mostly used to support business strategy and R&D, as input to product or process design, in education and for labelling or product declarations (Curran 1996). LCA also plays crucial roles in environmental impact assessment, integrated waste management and pollution studies; and was therefore introduced as one of the components in the curriculum at Symbiosis.

3.7 Environment, Health and Safety

With the establishment of US Environmental Protection Agency (EPA) in 1970 that later followed the emergence of Environment Health and Safety (EHS) as one of the important functions of the broad Environmental Management discipline around 1990s. With the enactment of Occupational Safety and Health Act of 1970, the importance of workplace safety and occupational health grew with time. Eventually, the companies formulated systematic ways and approaches of complying with EHS regulations. With the advancement of data technology management the things became easier for the companies to analyse its operations. During last decade of twentieth century, new management leaders in the form of 'EHS Manager' were created by the corporations to merge oversight of EHS. These leaders in their new role, who began their profession in one of the three sub-disciplines, created arrangements to drive EHS improvement across all operations.

Today, the organisations that seek to be better environmental stewards invest in strong environmental, health and safety management programmes, known as EHS. From an environmental facet, EHS is about crafting a systematic approach to manage waste, complying with environmental regulations and reducing the organisations carbon footprint. Efficient EHS programs in organisations also addresses ergonomics, air quality, hazardous and toxic waste and several other aspects of workplace safety that could affect the health and well-being of employees (MacLean 2003).

With the dawn of sustainability in 21st century, skills and experience of EHS managers have become more crucial than ever before. The EHS leaders are increasingly accountable for designing and executing strategies to take companies beyond compliance by way of reducing energy use, water management, implementing 3R principles to name a few. Some of the other critical initiatives taken by the EHS leaders include—evolving formal sustainability program; establishing corporate EHS standards and practices, integrating EHS values and practice across business; public disclosures; responding to stakeholder's inquiries; operating with supply chain; Global auditing; and ensuring good housekeeping at workplaces world across (Brown and Larson 1998; Patil and Kulkarni 2008).

In India, health and safety aspects have been long guided by the three main regulations viz. The Factories Act, 1948; The Mines Act, 1952; and The Dock Workers (Safety, Health & Welfare) Act, 1986. Over the years, while there has been more focus on only two elements of EHS i.e. Health (human) and Safety (fire), the third element 'E' i.e. Environment has gone completely neglected. Even the industrial safety diplomas and many other postgraduate programs' run across India are only focussing on industrial accidents and fire aspects thereby completely lacking the holistic approach. In order to bring in robust holistic elements Symbiosis introduced this course in their Energy and Environment programme.

3.8 *Sectoral Studies*

During the evolution of the Symbiosis Programme on Energy and Environment, the need for flexibility in curriculum development was often cited as a response from students and industry alike. From a student perspective this was necessary to develop creativity and the desire to inculcate a sense of questioning and analysing key emerging environmental issues. From industry perspective, it was felt that since the sustainability sector was a dynamic process, it was necessary to create an innovative curriculum aimed at sectoral aspects of sustainability. Using the scope of geographical (region specific, trade blocs) and sectoral areas (climate change, Renewable energy, EIA, supply chain, water management, etc.) a course was created which enabled students to bring out their own ideas and concepts for undertaking mini projects. This approach has led to focused peer learning process in classroom and distinct from a regulated University based curriculum.

4 Conclusion

Post the United Nations Conference on Environment and Development (UNCED) at Stockholm, Sweden (1972) there have been several energy and environment related programmes and initiatives across the world. Similar such programmes received impetus in India in the mid-1980s in various central, state and deemed universities on the basis of either government related regulatory frameworks or environmental issues of concern. Assessment of the past two decades clearly shows that curriculum development in India in the broad discipline of environmental science and management got limited to a particular niche area because of lack of apt human resource and unavailability of necessary teaching aids. Contingent to the available expertise the environmental curriculum in the eventual period got conversant with either chemistry or biology (ecology) or geography or geology or engineering. For instance, when the environment curriculum was chemistry dominated, the programme lacked its visibility on other fronts. This created a weak foundation among the learners. All being the core science disciplines focus on the managerial, social and economic aspects were completely missing. Human resource produced after studying such curriculum completely lacked the holistic (sustainable) approach to resolve environmental problems. While the energy and environmental related programmes and courses across India were being imparted in large number of universities, there are very few examples in which the acquired knowledge was being put into practiced for developing sustainable/green organisations and campuses due to one or other reasons, mainly being financial (Rao et al. 2015) obligations.

The significance of environmental sustainability as the future path for corporate action is perhaps the need of the time. This is particularly very relevant in context of some of the most challenging and complex global environmental issues the world is facing in 21st century; and climate change is one such key issue (Pachauri and Resinger 2007).This indicates that business as usual scenario will not work and there is very little time for the world to take action in combating impacts of GHG emissions which are likely to increase in an exponential way (Meinhausen et al. 2009). The rising demand for energy and its consumption in order to achieve higher economic growth is cited as a key driver of higher GHG emission rates (IEA 2008). Rapid industrialisation, urbanisation, migration and population overgrowth has certainly taken its toll by causing environmental stress (Mukhopadhyay and Revi 2009) and are the clear indicators of imbalanced human environment and ecology. Building and integrating sustainability based education across academic institutions is the key and need of the hour. In order to help and build sustainable and low carbon economy, strategic involvement of not only the country governments is needed but also business and industry, academic institutions and civil society needs to work in close cooperation and forge meaningful partnerships for implementation.

References

Areendran, G., Rao, P., Krishna, R., Mazumdar, S., & Puri, K. (2013). Land-use/Land-cover change dynamic analysis in mining areas of Singrauli District in Madhya Pradesh, India. *Tropical Ecology, 54*(2), 239–250.

Bonnett, M. (2002). Education for sustainability as a frame of mind. *Environment Education Research, 8,* 9–20.

Brown, H., & Larson, T. (1998). Making business integration work: A survival strategy for EHS managers. *Environmental Quality Management, 7*(3), 1–8.

Christensen, L. J., Peirce, E., Hartman, L. P., Hoffman, W. M., & Carrier, J. (2007). Ethics, CSR, and sustainability education in the Financial Times top 50 global business schools: Baseline data and future research directions. *Journal of Business Ethics, 73*(4), 347–368.

Curran, M. A. (1996). *Environmental life-cycle assessment.* McGraw-Hill Professional Publishing.

Date-Huxtable, E., Ellem, G., & Roberts, T. (2013). The low carbon curriculum at the University of Newcastle, Australia. In S. Caeiro, W. Leal Filho, C. Jabbour, & U. M. Azeiteiro (Eds.), *Sustainability assessment tools in higher education institutions—Mapping trends and good practices around the world* (Part IV, pp. 345–357). Switzerland: Springer International Publishing.

Dempsey, N., Bramley, G., Power, S. & Brown, C. (2009). The social dimension of sustainable development: Defining urban social sustainability. *Sustainable Development*. Retrieved March 24, 2014, from http://www3.interscience.wiley.com/journal/122406493/abstract.

Ehrenfeld, J., & Gertler, N. (1997). Industrial ecology in practice—The evolution of interdependence at Kalundborg. *Journal of Industrial Ecology, 1*(1), 67–80.

Ehrenfield, J. (1997). Industrial ecology: A framework for product and process design. *Journal of Cleaner Production, 5*(1–2), 87–95.

Finnveden, G., Hauschild, M., Ekvall, T., Guinee, J., Heijungs, R., Hellweg, S., et al. (2009). Recent developments in life cycle assessments. *Journal of Environmental Management, 91*(1), 1–21.

Garner, A., & Keoleian, G. A. (1995). *Industrial ecology: an introduction*. Ann Arbor, MI: National Pollution Prevention Center for Higher Education, University of Michigan.

Godemann, J., Herzig, C., Moon, J. & Powell, A. (2011). *Integrating sustainability into business schools—Analysis of 100 UN PRME sharing information on progress (SIP)* Reports No. 58-2011 ICCSR Research Paper Series, (J. Moon, Ed.). Research Paper Series, International Centre for Corporate Social Responsibility.

Heeres, R. R., De Vermeulen, W. J. V., & Walle, F. B. (2004). Eco-industrial park initiatives in the USA and the Netherlands: First lessons. *Journal of Cleaner Production, 12,* 985–995.

Hubbard, S. S., & Bowe, S. A. (2010). A gate-to-gate life-cycle inventory of solid hardwood flooring in the Eastern United States. *Wood and Fiber Science, 42*(3), 79–89.

Jelinski, L. W., Graedel, T. E., Laudise, R. A., McCall, D. W., & Patel, C. K. (1992). Industrial ecology: concepts and approaches. *Proceedings of the National Academy of Sciences, 89*(3), 793–797.

Leal Filho, W. (2000). Dealing with misconceptions on the concept of sustainability. *International Journal of Sustainability in Higher Education, 191,* 9–19.

Leal Filho, W. (2011). About the role of universities and their contribution to sustainable development. *Higher Education Policy, 24,* 427–438.

Leal Filho, W. (Ed.). (2012). *Sustainable development at universities: New horizons series: Umweltbildung, Umweltkommunikation und Nachhaltigkeit/Environmental Education, Communication and Sustainability* (Vol. 34, pp. 994). Frankfurt, Peter Lang Scientific Publishers.

Lozano, R., & Watson, M. K. (2013). Assessing sustainability in University curricula: Case studies from the University of Leeds and the Georgia Institute of Technology. In S. Caeiro, W. L. Filho, C. Jabbour, & U. M. Azeiteiro (Eds.), *Sustainability assessment tools in higher education institutions—Mapping trends and good practices around the World, Part IV* (pp. 359–373). Switzerland: Springer International Publishing.

Lozano, R. (2011). The state of sustainability reporting in universities. *International Journal for Sustainability in Higher Education, 12*(1), 67–78.

MacLean, R. (2003). Superior environmental, health, and safety performance: What is it? *Environmental Quality Management, 13*(2), 13–20.

Madeira, A. C., Carravilla, M. A., Oliveira, J. F., & Costa, C. A. V. (2011). A methodology for sustainability evaluation and reporting in higher education institutions. *Higher Education Policy, 24,* 459–479.

Matten, D., & Moon, J. (2004). Corporate social responsibility education in Europe. *Journal of Business Ethics, 54,* 323–337.

Meinhausen, M., Meinhausen, N., Hare, W., Raper, S.C.B., Frieler, K., Knutti, R., et al. (2009). Green house gas emission targets for limiting global warming to 2 °C. *Nature, 458,* 1158–1163.

Mitchell, R. (2011). Sustaining change on a Canadian campus: Preparing Brock University for a sustainability audit. *International Journal of Sustainability in Higher Education, 12*(1), 7–21.

Mukhopadhyay, P., & Revi, A. (2009). Keeping India's economic engine going: Climate change and the urbanisation question. *Economic and Political Weekly, 54*(31), 59–70.

Naeem, M., & Neal, M. (2012). Sustainability in business education in the Asia Pacific region: A snapshot of the situation. *International Journal of Sustainability in Higher Education, 13*(1), 60–71.

Pachauri, R. K., & Reisinger, A. (2007). *Contribution of working groups I, II and III to the fourth assessment report of the intergovernmental panel on climate change* (p. 104). Geneva, Switzerland: Core Writing Team, IPCC.

Park, J., Sarkar, R., & Bunch, R. (2012). Sustainability and management education in China and India: Enabling a global green economic transition. *Business Leadership Review, 9*(1), 1–15.

Patil, Y., & Rao, P. (2015). Industrial waste management in the era of climate change—A smart sustainable model based on utilization of active and passive biomass. In Walter Leal Filho (Ed.), *Handbook on climate change adaptation* (pp. 2079–2092). Berlin, Germany: Springer.

Patil, Y. B., & Kulkarni, A. R. (2008). Environmental sensitivity and management of toxic chemical waste in mining industry with special reference to cyanide. *High performing organizations: Needs and challenges* (pp. 183–196). Part D: Tata McGraw Hill Publications.

Patil, Y. B. (2012). Development of an innovative low-cost industrial waste treatment technology for resource conservation—A case study with gold-cyanide emanated from SMEs. *Procedia - Social and Behavioral Sciences, 37,* 379–388.

Porter, M. (1990, March-April). The competitive advantages of Notions.*Harvard Business Review.*

Rao, P., Patil, Y., Bhat, V., & Kulkarni, S. (2015). Environment and social sustainability approaches in policy and practice: A case study of symbiosis International University, Pune, India. In Leal Filho, W. (Ed.), *Transformative approaches to sustainable development at Universities, working across disciplines* (pp. 471–488). Springer International Publishing.

Rao, P., Patil, Y., & Gupte, R. (2013) Education for sustainable development: Trends in Indian business schools and Universities in a post liberalization era. In S. Caeiro, W. Leal Filho, C. Jabbour, & U.M. Azeiteiro (Eds.), *Sustainability assessment tools in higher education Institutions—Mapping trends and good practices around the world* (Part IV, pp. 417–432). Switzerland: Springer International Publishing.

Rao, P., & Yogesh, Patil (Eds.). (2016). *Reconsidering the impact of climate change on global water supply, use, and management* (p. 430). USA: IGI Global Publishers.

Saini, S., Rao, P., & Patil, Y. (2012). City based analysis of MSW to energy generation in India, calculation of state-wise potential and tariff comparison with EU. *Procedia - Social and Behavioral Sciences, 37,* 407–416.

Saqib, M., Sehgal, R., & Pamlin, D. (2007). Report on Indian companies in the 21st century (p. 66). WWF Publications.

Speller, P. (1992). Building partnerships for sustainable development: Federal University of Mato Grosso. *Higher Education Policy, 5*(1), 31–32.

Sterling, S. (2001). *Sustainable education: Re-visioning learning and change*. Green Books, The Schumacher Society, Schumacher Briefing No. 6, p 94.

Stubbs, W., & Cocklin, C. (2008). Teaching sustainability to business students: shifting mindsets. *International Journal of Sustainabilityin Higher Education, 9*(3), 206–221.

Tikoo, R. (2009) Management education gets an environmental edge. *The Financial Express.* May 22, 2009.

van Kleef, J. A. G., & Roome, N. J. (2007). Developing capabilities and competence for sustainable business management as innovation: A research agenda. *Journal of Cleaner Production, 15*, 38–51.

Wiek, A., Withycombe, L., & Redman, C. L. (2011). Key competencies in sustainability: a reference framework for academic program development. *Sustainability Science, 6,* 203–218.

World Bank. (2000). *Greening industry: new roles for communities, markets and governments.* World Bank Policy Research Report

Thinking About Sustainability: Issues and Themes for College Students

Gregory D. Bothun

Abstract This contribution discusses many of the themes that the author has used in various Environmental Studies courses at the University of Oregon. While this collection of themes could be viewed as eclectic, they are well rooted in what is happening in the real world. In cases of data, as this article was written December 2016, we have tried to find and use the most current data to describe some situation. This is quite important because in the arenas of energy generation, resource consumption and climate change, the real world strongly evolves and can change dramatically from what the situation looked like just a few years ago.

The material discussed here and its contexts are not likely to be found together in the same place in any existing text on the overall issue of sustainability. The themes chosen here are to remind the reader, at all times, of the big picture of sustainability within the context of over consumption of planetary resources and the subsequent change in the Earth's climate system. This material is organized as follows:

- Section 1 is a fairly large scale overview into the origins of the idea of sustainability and its initial connection to sustainable development as its main manifestation. In this section we (a) critically analyze previously published "definitions" of sustainability and sustainable practices; (b) introduce William Cronon's essay on *The Trouble with Wilderness* as a philosophy that embodies best principles for sustainability; (c) use the evidence for accelerating accumulation of atmospheric CO_2 as the primary example that we are not globally acting in a sustainable manner; (d) use the data on the rapidly increasing amount of container traffic in the global distribution of goods as the prime example of escalating consumption which ultimately drives climate change.
- In Sect. 2 we consider sustainability as a cycle similar to a life cycle product analysis. All cycles are characterized by two things (a) material exchange rates between various reservoirs and (b) the existence of buffers in the system that can

G.D. Bothun (✉)
Department of Physics, University of Oregon, Oregon, USA
e-mail: bigmoo@gmail.com

J.P. Davim (ed.), *Curricula for Sustainability in Higher Education*,
Management and Industrial Engineering, DOI 10.1007/978-3-319-56505-7_6

store some material over various timescales. In this way, a system can be out of equilibrium on either short spatial or temporal timescales, but on average equilibrium is always maintained. We discuss the hydrological cycle as a good example of a system in natural equilibrium. We then discuss the Nitrogen cycle as an example where human intervention has strongly altered one of the material exchange pathway rates. We then discuss 4 real world product cycles in this context of natural cycles. These are (a) the life cycle of mobile phones in terms of recycling versus reuse; (b) the energy cycle costs associated with producing and consuming a cheeseburger as this provides a good example of net energy loss; (c) the use of water in the overall cycle of producing a 0.5 L bottle of Coca Cola and its associated packing; (d) the coming issue associated with the need for developing a much better recycling/reuse system for lithium ion batteries.

- In Sect. 3 we introduce the concept of business as usual (BAU) in the context of our continuing desire to exploit fossil fuel reserves. We begin that section with a novel conceptual story about cultural mixing between various tribes on two different kinds of planetary surfaces. The evolution of these tribes will be driven by their accessibility to resources. On one planet, resources are easily accessible while on the other planet they are not. This consideration gives rise to the concept of the **sustainability radius** which defines a geographic region in which a specific culture (tribe) can live on the many generations timescale. This section ultimately raises the question of whether or not it's inevitable that the sustainability radius has to expand to the radius of the planet.
- Section 4 describes the idea of sustainability versus dominance over the Earth's resources in a historical context. Much is discussed about the rise of the Mechanical Philosophy, primarily through the works of Descartes, which ultimately treats that Earth as nothing more than a machine and establishes the man is distinct from nature and therefore entitled to dominant it. This Mechanical view of the world is firmly in place before the industrialized revolution and therefore it seems clear that the concepts of sustainable harvesting of Earth resources via machines (e.g. the steam engine) are nowhere to be found. The manifestation of such unbridled dominance over nature is discussed in the real world contexts of coal production (primarily in the UK), coal befouled cities in the US, and the ultimate rise of conspicuous consumption during the post-World War II era. In historical context, sustainability appears now to be a reaction to the 150 year period form 1850–2000 of our industrialized behavior.
- Section 5 discusses our recent movements towards the concept of "just sustainability" as most strongly articulated by Julian Agyeman (2005). In this concept, issues of social justice and overall equity are viewed to be more prominent that sustainable development in the economic sense. The argument made here (also by others) as that our collective value system needs be based on things other than just increased prosperity through economic means. Issues of fairness, environmental justice, dignity and global equity should become more important if we are to manage our planet better. This section discusses in detail the idea of living within ecosystem limits and makes use of the 2016 Living Planet report for a focused discussion on our global ecological footprint.

- In Sect. 6 we present a brief discussion of climate change that is driven primarily by our systematic heating of the oceans through industrialized processes and their associated waste heat. This section shows and discussed the most recent data on the issue of Arctic warming, rapid loss of Arctic Ocean sea ice, and subsequent release of methane as permafrost (mostly in Siberia) begins to melt. We are on a serious course leading to an ice free Arctic ocean. The loss would significantly change the albedo of the Earth which would substantially change the Earth's overall energy balance. This likely represents the one tipping point in the climate system in which it is not possible to recover from. We attribute this course entirely too escalating global consumption as the primary driver.
- We present some concluding remarks in Sect. 7 and map some of the well-established business reasons for not changing business culture onto consumer habits that need to change to illustrate the overall difficulty of shifting away from consumption drive business as usual practices to more sustainable practices.

This contribution also contains two Appendices. Appendix 1 presents a brief overview and links to the Zero Waste program at the University of Oregon as a good example of quantitative sustainability achieved on the local scale. Appendix II contains some examples of possible student exercises that are related to the material we present here.

1 Introduction and Overview

Sustainability: What is it? Why is it good public policy? How does one measure if sustainability has been achieved so that the policy can be evaluated? The first two of these questions have been discussed at length in a variety of texts and other sources; while we will touch upon these two topics in this presentation, the bulk is concerned with the later, much under discussed issue, of the **physical** aspects of sustainability. In general, while sustainability is treated and taught as some **qualitative** set of conditions and ideals, the actual footprint of global (increasing) consumption on the function of the Earth and its resource availability is very much a **quantitative** issue. Students that desire to increase their knowledge of sustainability need to be education in how systems work through pathway flows, buffers, and end use. These variables are all measurable and can quantitatively define how a system works, and using these measurements as a guide, system flows can be improved and perhaps even optimized. However, all too often the system is ignored and issues of sustainability become narrowly focused on some local problem that may or may not be relevant to the sustainability of the system. It is the focus of this chapter to consistently discuss the issues of sustainability in the larger system context.

Historically, the sustainability concept seems to be intertwined with the more narrow and specific goal of 'sustainable development'. Indeed, this is the central concept emphasized by the impactful 1987 Brundtland Report *Our Common Future* which defines sustainable development as "*development that meets the needs of the present without compromising the ability of future generations to meet their own needs*,". Well, this is a nice, friendly public policy statement that should offend nobody and be embraced by policy makers and planners as a sensible way of planning for the future. However, there are two quantitative elements of this qualitative statement that now lie at the issue of sustainable resource management: (1) the "needs of the present" require a quantitative measure—for instance is such a need 1000 gallons of gasoline per year per person, etc.; (2) "compromising the ability of the future"—this is a physically implausible statement to even make when it is known at the time that resources are finite. At some point in the future, global consumption obviously compromises the resource availably for that future generation—this is the natural outcome of consumer consumption of finite planetary resources, unless resources can be recycled on the same time scale that they are used. This is not physically possible to due, given the existence of buffers (see more below) in any system. And so the Brundtland report (see also Middleton and O' Keefe 2001) establishes the qualitative notion that sustainable development can be achieved so therefore there will be no future compromise. This is a wishful thinking statement made at the time since now, 30 years later, it seems like an absurd statement as compromise is beginning to happen in this current "future" generation in a lot of areas.

A point of emphasis in this chapter on this issue will involve the future availability of rare earth elements and the continuing support of improving technology. In other words, does sustainable development really involve making sure that the continual billions of consumers have unfettered access to the latest generation Iphone? If previous Iphones could be **fully** recycled to make new ones, then, of course, this would be sustainable. But this is not the case and in some instances it is known that recycling the components takes more energy than attempts to repurpose the components. Furthermore, according to Lisa Jackson, the head of environmental affairs at Apple, "*Apple shreds their devices to avoid having fake Apple products appearing on the secondary market*" (retrieved from this report: https://9to5mac.com/2016/02/17/recycled-iphone-what-happens/). This indicates that corporate policy is getting in the way of doing the most sustainable thing with this product. We are not attempting to single out Apple here but just to illuminate the general concept that (a) recycling does have an energy cost (this is detailed later in this chapter) and (b) no product can be fully recycled in the manner suggested by the replica machine on Star Trek. All recycling leads to some degradation of base resource and those degradations have to be replaced by new resource and the amount of that new resource is dwindling. We will encounter this issue in the section below on lithium batteries.

Some manifestation of this notion of diminishing availability has been noted before through various objects of the sustainable development mandate of the Brundtland report as this report implicitly assumes that material needs can be met within ecosystem limits (for more on this see Jacobs 1999; Bourke and Meppem 2000; Gunder 2006; Connelly 2007). In addition, the Brundtland report was published long before the true global scale of our consumption was known. Indeed, on this global scale, the boundary conditions of achieving sustainability are daunting: the world is inhabited by diverse populations with highly variant value systems and perceptions of nature. Hence the concepts of "development" and "needs" very much have a cultural context. This context has been explored in Larrain et al. (2003) who introduce the concept of the 'dignity line' as cultural specification of the minimum level of material consumption required to maintain life with dignity. So, does entitled access to the latest generation I-phone enhance our dignity? Evidence to date suggests that humans collectively behave in a manner that simply consumes the Earth (in a highly culturally differential manner) and the ideas of dignity and minimal consumption do not factor into our daily behavior. Indeed, the overall hypocritical mismatch between our daily behavior and our stated values has been well described in William Cronon's 1995 essay *The Trouble with Wilderness* (see student exercise at the end of this chapter). Cronon's basic point is that we culturally define wilderness as a distant pristine place that we, as steadfast environmentalists, should protect. Therefore, when we are physically in the wilderness our behavior changes (perhaps we don't even take the Iphone with us). Cronon argues that we need to adopt an ancient value system based on the simple notion that "*wilderness is everywhere*" and under that value system better behavior will occur.

To date, very little progress has been made on sustainability on the global scale despite the large amount of discussion, conferences and journal articles devoted to need to develop much better sustainability practices across all sectors. An early example of this lack of progress involves the 1972 publication the book, *Limits to Growth* (Meadows et al. 1972). Ultimately this did little to evoke much action, especially from those in government and corporate sectors bent on economic growth as the principle measure of prosperity. While it can be argued that prosperity defined in a manner which ignores issues of dignity and equity is a profoundly western view of the world, it seems clear that this view is the principle driver in our evolution as a consumptive species.

Additional evidence of this lack of progress in achieving sustainability is available through several key quantitative indicators, one of which is the accelerating output of CO_2 by the collective consumption actions of the planet. In Table 1 below, we summarize the decadal averages of CO_2 increasing from the recorded measurements at the Mauna Loa Observatory from 1958 through December of 2016:

Table 1 CO_2 emission levels per decade

Decade	Level mid-decade	Mid decadal average increase	Average annual PPM increase in the decade
1960–1969	July 1965: 320.5		0.86 ± 0.50
1970–1979	July 1975: 331.17	10.67	1.22 ± 0.62
1980–1989	July 1985: 345.8	14.63	1.61 ± 0.53
1990–1999	July 1995: 361.11	15.31	1.54 ± 0.74
2000–2009	July 2005: 379.85	18.74	1.90 ± 0.56
2010–2016	July 2015: 400.76	20.91	2.43 ± 0.66

The data clearly show that starting since the year 2000; average annual CO_2 emissions are 25% higher than the values for the previous decade. For the most recent decade, the annual CO_2 emissions are almost a factor of 2 higher than they were over the period 1960–2000. So, CO_2 emissions from our combined consumptive activities are **accelerating** despite the existence of the many symbolic international agreements and conferences (e.g. Paris 2015) aimed at limiting CO_2. Now it is important to point out that even if CO_2 emissions were to stabilize that does not mean that there would be an immediate stabilization of atmospheric CO_2 levels. This is because current the natural ocean and land sinks to reabsorb atmospheric CO_2 are only able to offset 50% of current global emissions. So the remaining 50% still accumulates in the atmosphere. Moreover, due to rising ocean temperatures, the oceans ability to mix out atmospheric CO_2 is declining. To fully stabilize atmospheric CO_2 at some level would require an immediate 50% reduction in current emission levels, quite contrary to the data trends shown in Table 1. However, a sliver of good news has recently emerged as CO_2 emissions related to energy (transportation, electricity, heat) generation appear to have flattened (Fig. 1).

Perhaps the best proxy for our increasing consumption, despite our supposed increase in environmental awareness, is shown in Fig. 2 which plots the number of Twenty Foot Equivalent (TEUs) containers of consumer goods shipped all over the world on an annual basis over the time period 1980-through 2013. While the 2009 global economic meltdown is clearly revealed in this data (denoted by the red arrow in Fig. 2a), the consumption rebound was swift and has continued throughout 2010–2013. Overall there has been approximately a tenfold increase in container

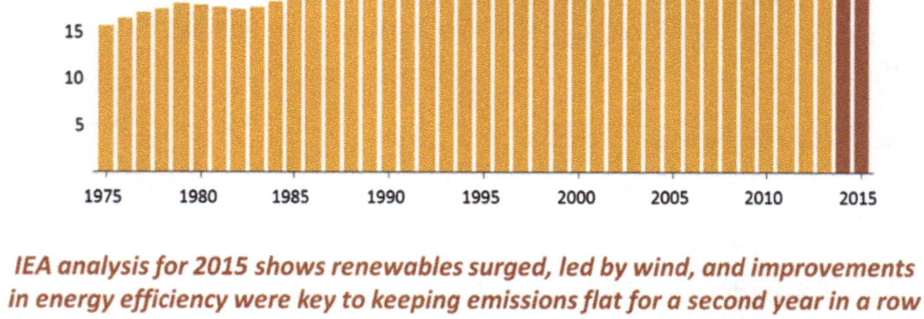

Fig. 1 Date from International Energy Agency 2015 Annual report

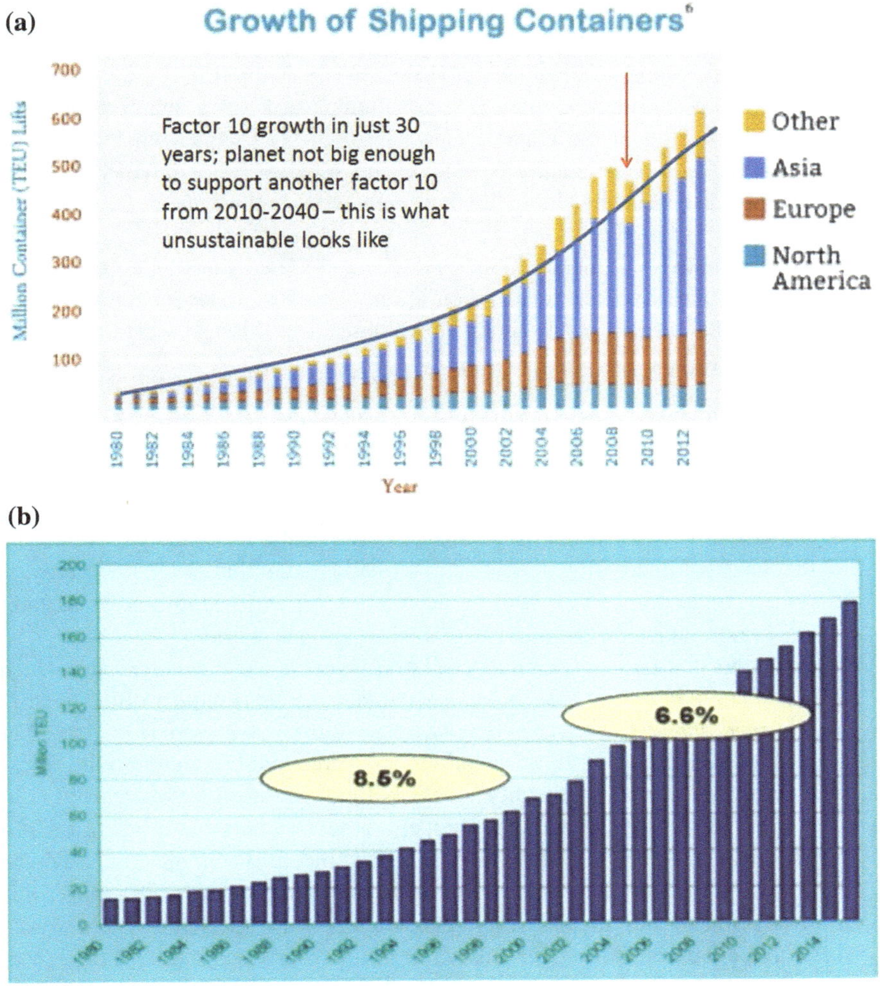

Fig. 2 **a** World shipping container traffic (data from Drewry Shipping Consultants). **b** Updated from Fig. 2a to includes 2014–2016 data

traffic during the 30 year period of 1980–2010. This is the quantitative consumption scale that sustainability initiatives must address and it's quite clear that in the real world, humanity is currently living in the most unsustainable time in history. Somewhat more recent data on container traffic is shown in Fig. 2b. The growth rate as defined by the period 1980–200 is 8.6% per year, which has lowered slightly to 6.6%. Still that means the overall magnitude of goods shipped is increasingly large and this is necessitating drastic measures such as (1) building even larger ships; as of 2015 these ships could carry 19224 TEU—in 2011 the largest ships carried about 13000 TEU; (2) the escalating of build our or world container ports—in 2014 the largest port at Shanghai processed 35.29 Million TEUs for that year (data from World Shipping Council -worldshipping.org); (c) a tentative plane to make a larger version of the Panama Canal through Nicaragua with most of the traffic spent on Lake Nicaragua. The scale of this escalation of container traffic so that consumers can get their global goods should serve as the principle quantitative reality of our consumption scale.

Now that we have established, through data, that the real world of consumption, resource usage and atmospheric pollution are accelerating, it seems appropriate to revisit various definitions of sustainable development in this real world context:

1. From the World Commission on Environment and Development: Sustainable Development is [sic] *A process of change in which the exploitation of resources, the direction of investments, the orientation of technological development and institutional change are all in harmony and enhance both current and future potential to meet human needs and aspirations.* In the real world context it seems safe to say that investments and technological development currently lie at the heart of increasing resource exploitation and that such a synergy is extremely inharmonious as well as destructive. So this definition of sustainability seems rather far from our activities in the real world.
2. In Forum for the Future: *Sustainable Development is a dynamic process which enables people to realize their potential and improve their quality of life in ways which simultaneously protect and enhance the earth's life support systems.* For starters, this rendering of the sustainability concept would seem to require a new mechanism that improves quality of life without the need for increasing consumption. To initiate this new requirement, a new value system has to replace the old value system based largely on economic decision making. Such an evolution of value system indeed would be a dynamic process. Of course, this reformulation of the sustainability mandate offers no mechanism that would drive this evolution. Moreover, according to the most recent (2016) edition of The Living Planet Report (http://wwf.panda.org/about_our_earth/all_publications/lpr_2016/) there was a 58% overall decline in global vertebrate species population over the period 1970–2012, which hardly supports the notion that we are "enhancing" the earth's life support systems.
3. From Making London Work by Forum for the Future's Sustainable Wealth London Project: *In essence sustainable development is about five key principles: quality of life; fairness and equity; participation and partnership; care for the environment and respect for ecological constraints—recognizing there are 'environmental limits'; and thought for the future and the precautionary principle.* This is a very egalitarian view which calls for an integrated process of sustainable development. Much of this view would seem to require some institutional evolution where policies based on fairness, equity and

environmental care should have precedence over economic ones. Indeed, such an approach might work for small developed European countries but this approach would remain local and not global. In addition, for countries that have basically used up most of their resources, and thus require on importation from the rest of the world, it's fairly easy to talk about 'environmental limits'. What is needed instead, from this policy framework, is a mechanism that can evaluate and monitor these environmental limits on a global scale. The litmus test of these global environmental limits lies in our treatment of the world's remaining freshwater resources (Wyman 2013). Recent reports suggest global depletion of many freshwater resources by 2040 (see http://www.planetexperts.com/new-studies-freshwater-resources-will-depleted-2040/). While these two recent reports seem to be overly alarming, there is no question that on the longer timescale (out to 2080) increasing using of water for electricity generation and agriculture combined with the drying out of the Tibetan plateau from global climate change (Ren et al. 2004) shows that human management of the planet is most definitely on a trajectory to deny most of its citizen's access to freshwater.

4. From the World Commission on Environment and Development 1996 Version: *The environment must be protected ... to preserve essential ecosystem functions and to provide the wellbeing of future generations; environmental and economic policy must be integrated; the goal of policy should be an improvement in the overall quality of life, not just income growth; poverty must be ended and resources distributed more equally; and all sections of society must be involved in decision making.* Of all the alternative definitions offered here, this one contains two key points, missing from most other views of sustainability: (1) *resources distributed more equally*—this seems to be a vital need for achieving sustainability but currently we are rather far from this ideal (see for instance Armstrong 2011). The 2008 World Bank report showed that for 2005, the world share of consumption can be broken down as follows:

   - 75.6% is consumed by the Worlds' richest 20%
   - 21.9% is consumed by the middle 60%
   - 1.5% is consumed by the world's poorest 29%

   This is anything but equal and although more recent data do not yet seem to be available (this is a difficult data gathering process to make such estimates) it seems rather unlikely that equity has improved over the last 10 years; (2) *not just income growth*—this is consistent with the previously mentioned needs of changing the value system. As long as the personal perception of prosperity is related solely to income issues, which ultimately drive consumption, then no move to sustainability is possible. This is one of the major challenges for the more developed countries of the world—how to enlighten their citizens that the drive for steady personal income growth is now less important than considerations which raise the probability of a livable world for future generations.

5. From the Dorset Education for Sustainability Network: *We cannot just add sustainable development to our current list of things to do but must learn to*

integrate the concepts into everything that we do. Exactly. This is what Cronon (1995) has urged us to do—learn that wilderness is everywhere. Currently, we do treat the issue of sustainability as an item on a task list, with a remark saying something like "yeah, let's work on this today".

6. From the Learning for a Sustainable Future Teacher Centre: *A sustainable future is one in which a healthy environment, economic prosperity and social justice are pursued simultaneously to ensure the well-being and quality of life of present and future generations.* This formulation embodies an idea which has increasing momentum—namely social justice. This often goes in the form of global justice (see Pianta and Marchetti 2007) or climate justice. More practically this idea of social justice has been explained in terms of global public health or nutrition (and currently there are several PHD programs in the US that are centered on these issues). This overall idea the economic prosperity and social justice can be commensurate goals does represent a new way of organizing a consumptive society around a set of more mutual benefits than individual personal gain. As such, this seems like a very promising way to promote sustainability and to measure if any of its goals are being reached.
7. From Worster 1993: *The first and perhaps most difficult problem, one that seldom gets addressed, is the time frame ... Is a sustainable society one that endures for a decade, a human lifetime, or a thousand years?* This issue is directly addressed near the end of this chapter. As discussed earlier, sustainability is largely a qualitative term, and sustainable development implicitly carries with it the concept growth and the ideal that such growth can be managed in a more beneficial way than has been done in the past. However, that ideal is certainly being challenged and there is no evidence to suggest we have obtained it or will obtain it in the near future. Worster's 1993 question is really about extracting resources from the Earth on a timescale that is closer to some of the natural timescales in the Earth system (see the Nitrogen Cycle example in Sect. 2). Ever since the industrial revolution, we have turned the resource production crank as fast as possible to not only maintain our growth trajectory but in fact to seemingly grow at the maximum possible rate. For example, we will likely have used up the Earth's accessible fossil fuel source in 200 years. Issues like climate change may have been less threatening had we decided to use up this reserve in 1000 years, instead of this more compressed time frame. According to Ernst (2002) if we want to move towards a more sustainable level of earth resource consumption we must (1) develop very efficient new extraction techniques, (2) improve our efficiency of materials recycling and (3) find a new ubiquitous and inexpensive source of energy to power 1 and 2.

In sum, the academic view of the term "sustainability" has a wide ranging set of definitions and concepts although its roots in the original vision of "sustainable development" have not withered. Efforts to achieve sustainability via the route of "consuming less" are clearly not happening in the face of real world data showing that global consumption and associated atmospheric CO_2 accumulation are accelerating. This indicates that whatever local (or personal) movement there is towards sustainable behavior, as a global species we are not practicing that with respect to our one and only planet.

2 Sustainability as a System

A system is one in which the various components are consistent interacting with other through a network of exchange pathways and reservoirs. Individual material (or atoms) can move through this network and appear in different reservoirs at different times, but ultimately all parts of the system have interacted with the material. As we will see, it is very useful to look at sustainability in the context of such a system. To initiate this view, we will describe to important natural cycles—the hydrological cycle (Fig. 3a) and the nitrogen cycle (Fig. 3b). The hydrological cycle provides an excellent example of pathways, buffers and exchange rates. At any given movement the system may be out of equilibrium (both spatially and temporarily) but on some timescale the equilibrium of the system is maintained meaning that none of the reservoirs systematically increases or decreases on that time scale. The Nitrogen cycle is included here because it provides a direct example of the human manipulation of natural exchange rates thus allowing excess nitrogen to be deposited in reservoirs in a **non-equilibrium** manner.

This depiction shows the basic process of flows and reservoirs for the fresh water resource that gives rise to equilibrium. The basic process is this:

1. Since the oceans are not growing in volume, that means there is no net loss of gain of water. Therefore the amount of water that leaves the ocean must equal that which enters the ocean.
2. There are 413 units of water that leave the ocean via evaporation. Of those, 373 falls directly back into the ocean as precipitation meaning that 40 units temporarily are relocated away from their ocean reservoir. Now, if there were no return channels back to the oceans, then the ocean reservoir would be in a net decline.
3. The return channels to the ocean are shown above with the net aggregate of 40 units of surface flow returning to the ocean thus maintaining its equilibrium.
4. The timescales of the return of the 40 units to the ocean are considerably longer than the timescale associated with direct oceanic precipitation. Once the water transport gets to land it can be temporarily stored in various buffers, such as snowfall, lakes, in vegetation, soil moisture or as ground water. But each of this buffers will eventually release that water and return it to the oceans thus **maintaining** equilibrium.

Thus for the hydrological cycle, none of the resource (water) is ever lost to the system. The overall effects of climate change, however, will manifest themselves by changes in the various exchange rates. But this has happened many times before as

(a)

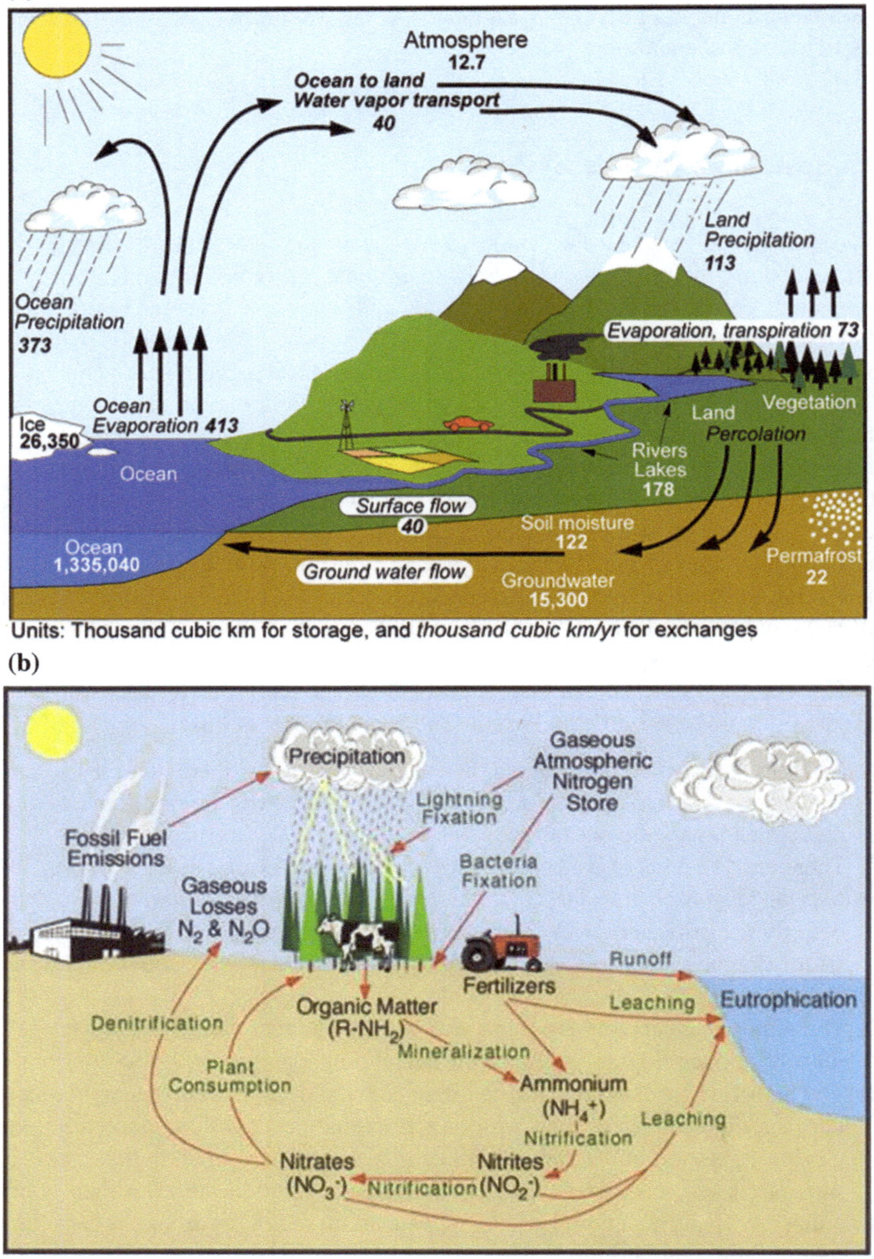

Fig. 3 **a** Representation of the hydrological cycle by Trenberth et al. (2007). **b** Representation of the Nitrogen fixation and release cycle

past Ice Ages have had major effects on the hydrological cycle rates and the ocean volume has reduced because the land ice volume has increased, but still the cycle is maintained and equilibrium (on long time scales) is also maintained. So this is a good framework for describing the flow of consumption and consumer products and we will provide such an analysis later in this chapter.

We now turn to the Nitrogen (N_2) cycle where the tractor represents human intervention:

The various processes illustrated here are the following:

1. Our atmosphere is mostly made of N_2 but that nitrogen cannot be used directly by plants but has to be deposited into the soil by other processes. These processes involve the fixation of N_2 into the soil through bacteria removal of atmospheric N_2 and through the combined actions of atmospheric lightning and precipitation that can also remove atmospheric N_2 and deposit this in the soil (once the N_2 is in the soil it combines with other elements to make Ammonium which is what plants use as a nutrient).
2. In an equilibrium system, the amount of N_2 returned to the atmosphere is equal to the amount that is removed. The principle way that N_2 is returned to the atmosphere is through the action of de-nitrifying bacteria in the soil.
3. However, some N_2 will be part of the surface runoff associated with atmospheric precipitation and that will end up in an unwanted reservoir, namely coastal estuaries. When N_2 is chemically introduced into ocean waters, it tends to remove some of the oxygen from those waters (this is called Eutrophication) adversely affecting the local shell fish population. That excess N_2 does eventually leach out back to the soil under the estuary to make its way back to the de-nitrifying bacteria.
4. Humans, however, directly interfere with this process due their large scale mining of atmospheric nitrogen for the purpose of fertilization. Most of this fertilization is immediately subject to surface runoff and thus we have very much increased the rate and deposit from surface runoff into ocean estuaries. In essence, humans have essentially mined atmospheric N_2 and dumped into the ocean estuaries at a much faster rate than would have occurred in nature. This is one of the best examples of human activity induced changes in a natural exchange rate with the adverse effect of moving a resource (in this case N_2) to a reservoir that it doesn't belong in (see Howarth and Marino 2006 for an extensive review).

Within the previously defined cycle framework it is now useful to look at an example for a consumable resource, in this case, the lifecycle of Mobile Phones (retrieved from http://www.unep.fr/shared/publications/pdf/DTIx1208xPA-LifeCycleApproach-Howbusinessusesit.pdf). The overall goal of this cycle would be to avoid the landfill box as that now represents disappearance of a resource and its loss to the overall system. In the business world, Fig. 4 is known as the "cradle to grave" process, where the ultimate goal is to preserve as much of the initial material (cradle) to recycle into new components thus avoiding the graveyard.

Fig. 4 Lifecycle of mobile phones

This cycle is a strictly quantitative cycle where various percentages or masses of components are maintained in the reuse cycle and not discarded. However, attempts to measure the reuse efficiency of individual components are generally not done, but a final assessment of remaining mass is all that is recorded. So when Apple claims that 85% of an I-phone is recycled, that just means that 85% of the total initial mass of the phone has been recycled. However, the remaining 15% is likely in the key components (rare earth elements) that actually make the device work. Hence in Fig. 4, after the collection circle, we see the recycling broken out into various components, batteries, handsets, and accessories. It is highly probably that the reuse rate of these components is highly variable and indeed most all of the battery material will end up in a landfill whereas the handset and accessory items can probably be directly reused once they are separated from the overall device.

In addition to material cycle costs, there is also energy costs associated with each component of some device. Here, we consider the cheeseburger as a device. When consumed, the cheeseburger supplies about 500 calories of energy to the consumer, but the total energy input as shown below is closer to 1750 calories. So it takes 3.5 times more energy to produce the cheeseburger than it does to consume so eating a cheeseburger is very energy intensive. In addition, this obviously represents a net negative energy flow and if most of that energy input is fossil fuel based, then cheeseburger consumption leads (indirectly) to climate change (discussed briefly at the end of this chapter).

Each of the energy components from cradle (birth of cow) to grave (consumer eating the cheeseburger) is shown here in boldface:

1. We have to grow and feed cows for the meat and the cheese, so that is the **agricultural** component.
2. We have to **process** the cow in order to extract beef and cheese patties.
3. We then have to **store** (freeze) these patties before they can be **transported** to the cheeseburger maker.
4. Finally, we have to **cook** the patties for the taste buds of the consumer.

The energy costs associated with this entire production cycle are summarized in the Table below:

Energy step	Calories used per cheeseburger	Percentage contribution to cheeseburger energy cycle
Agriculture	953	55
Processing	251	14
Storage	190	11
Transportation	160	9
Cooking	189	11
Total	1743	100%

So given this information, how would you make a more sustainable cheeseburger? Well, if you could go directly from processing to cooking (i.e. you have ON site cows and processing facilities—not practical or likely not even healthy) then you could eliminate the storage and transportation costs, but those are only 20% of the total anyway, so you still have net negative energy flow. This example also shows that, like most energy step analysis, the bulk of the energy costs occur in one step, in this case, the energy it takes to grow the cow to the point that it can be turned into cheeseburger patties. In principle, if consumers had all this energy information available for them in various consumer products, they could make better choices on which products to consume.

Another example of this kind of footprint analysis comes from a water assessment study performed by the Coca-Cola company in 2008 (**see** http://www.thecoca-colacompany.com/presscenter/TCCC_TNC_WaterFootprintAssessments.pdf.) Water resources can come from three sources:

- Blue water = direct fresh water from lakes, streams, springs
- Green water = water generally stored in the soil and available to crops
- Gray water = recycled water

Water use was accounted for both in the production of the various ingredients for Coca-Cola as well as the water use involved in making the various packaging components (bottles, labels, and packing materials). The assessment was made for and 0.5 L bottle of Coke where the bottle was made from the plastic polyethylene terephthalate (PET). As we will see in the breakout below, 35 L of water resource will be required to be this 0.5 L bottle of consumer product; 70 times more water is used than consumed.

The breakdown of water usage found in this study is the following:

Bottles, labels and packaging used 7 L	Ingredients use 28 L	Total Water Use
83% gray water	20% gray water	Grey = 12 L
13% green water	52% green water	Green = 15 L
4% blue water	28% blue water	Blue = 8 L

The above analysis reveals that if you want to lower the use of green and blue water in your overall product you need to analyze water use on the ingredients. This was done with the result that the production of sugar was the most significant component of water use as about 17 of the 23 L of green and blue water were used for the production of sugar beets. Hence use of a sugar substitute would make this produce much more sustainable.

We conclude this section with the important issue of batteries and recycling/reuse of batteries and their components. As of 2017, there is very large interest in the world wide lithium supply chain and availability particularly as it relates to improved availability for increasing electric vehicle production (see for instance Chung et al. 2015—https://energy.gov/sites/prod/files/2015/06/f23/Lithium-ion%20Battery%20CEMAC.pdf). There is also significant new research and support to develop batteries with higher energy density capacity. A particular recent example is the Battery 500 project (https://www.energy.gov/technologytransitions/articles/battery500-consortium-spark-ev-innovations-pacific-northwest-national) whose goal is to produce an energy density of 500 W hours per kg, which is 3 times higher than the current capacity in electric vehicles which ranges from 150–180 W hours per kg. In addition, the increasing prevalence of devices (laptops, phones, tables, drones, etc.) is putting further strain on available lithium supply chains is for the near future, the best choice for battery material remains lithium and the use of lithium metal as an anode. In view of this it becomes important to efficiently recycle the various lithium batteries components as devices and vehicles reach their end of life. At the moment, while we recycle lead-acid batteries, we do not yet in large quantities recycle lithium (ion, metal, polymer) batteries because their end of life will generally not be occurring until 8 to 10 years from now.

For alkaline batteries, materials are abundant and non-toxic so it is unclear if recycling versus disposal produces any net gain (Babiak and Boolish 2009). However, for automotive batteries, there are both environmental and economic benefits of recycling. Materials can be recovered from used batteries and repurposed meaning that less raw material needs to be extracted from the limited supplies in the ground. In addition, using recovered materials such as copper, nickel and cobalt avoids the mining and smelting of more sulfide ores. Automotive batteries, for the most part, remain the traditional lead-acid battery type. In the United States an astounding 99% of automotive lead acid batteries are recycled (see http://c.ymcdn.com/sites/batterycouncil.org/resource/resmgr/BCI_Recycling_Rate_Study_200.pdf). The basic process of component recycling of lead acid batteries works as follows:

1. Plastic is used to make battery cases and covers and this material is recovered and reused to make new cases and covers.
2. Lead and lead oxide components are directly re-used to make new grids and batteries.
3. The electrolyte solution in car batteries generally uses sodium sulfate mixed with water (this makes H_2SO_4 otherwise known as sulfuric acid). During the recycling process, sodium sulfate crystals are recovered and re-purposed in textiles, glass and detergent manufacturing. In other cases, the electrolyte can be neutralized and sent to a local water treatment plant or the used electrolyte can be chemically recharged and used in a new battery.

This relatively simple separation and re-use of the various components is highly successful for lead-acid car batteries and represents one of the most sustainable product life cycles in existence. For comparison, various recycling studies show that 62% of paper, 58% of aluminum cans, 40% of glass and 22% of solid waste are recycled (see Gaines 2014). Given that lithium-ion batteries are now being used in significant quantities for electric vehicles and that these battery packs are designed to last the life of the vehicle (10–15 years) then there is a coming need to develop a battery recycling system for them, similar to the highly successful model we already employ. In addition, lithium batteries used for electrical energy storage (see the Powerwall exercise at the end of the chapter) is also increasing. Hence, 10 years from now, there will likely be a relatively large mass of spent Li-ion batteries that need to be recycled, instead of ending up in a landfill or exported to other countries with less stringent environmental health regulations. Can the established model of lead-acid car battery recycling be modified and used for lithium based batteries?

One complication of recycling Li-Ion batteries is that their active materials are usually in the form of a power that coated onto each metal foil and these different materials need to be separated in any recycling process. Moreover, a Li-ion battery pack can have 100 s or even thousands of individual cells (like in a Tesla Vehicle) that are connected with various control and thermal management circuitry that need to be recovered mostly intact for efficient reuse. In addition, the materials for the cathode element of the anode Li-Ion (or lithium metal) varies by different manufacturer and can involve elements like Cobalt, Nickel, Manganese, and Aluminum. Hence, some method of separation for the cathode materials needs to be developed.

Finally, current small-scale recycling of lithium-ion batteries requires high temperature furnaces and the introduction of a slag-forming agent so that during the melting process of the recycled components the slag contains lithium, aluminum, silicon, calcium, iron and any other elements that were part of the cathode material (e.g. manganese). This is an energy intensive process and recycling of aluminum or lithium from the slag base material is currently not economical and not at all energy efficient. In addition, the entire recycling process can produce various toxic byproducts. Hence, unless there is significant technical evolution in the li-ion recycling process, in 10 years, there will be millions of kilograms of used Tesla automotive battery packs occupying various landfills. The biggest challenge at the moment seems to be developing a more energy-efficient way to recycle such a battery pack.

3 A Framework for Non-sustainable Behavior of Humans

A simple argument can be made, that prior to the industrial revolution, no matter what attitude that humanity or culture had towards nature, no real harm to nature could come from that attitude as we had not yet developed machines (e.g. the steam engine). Although attitude towards nature have high cultural variance, what matters to the Earth system is the collective behavior of humans. Usually, that collective behavior illuminates the prevailing attitude and value system. We now consider two possible but orthogonal principles that exist before the coming of the steam engine, regarding human's place in nature.

- Attitude 1: Humans are in a necessary partnership with nature and its resources.
- Attitude 2: Humans are special and separate from nature and therefore are entitled to dominate its resources.

Before discussing a historical approach to this dichotomy, it is instructive to discuss two specific hypothetical scenarios which automatically can give rise to either attitude 1 or 2. Figure 5 shows two possible patterns for the physical geography of a planetary surface and the subsequent evolution of resource usage on those surfaces. Planet A on the left shows a geography that consists of large (on the human scale) connected continental land masses separated by water. Planet B on the right shows a surface that consists of a large number of small islands separated by sufficient distance so that no one island can see the existence of another island over the radius of curvature of the planet. Each planetary surface has been subject to the same kind of bio-chemical-physical evolution so has produced the same kind of species and same kind of resource based. On each planet we create the two cultural views defined by attitudes 1 and 2 above. Cultures that are driven by attitude 1 are symbolically shown by the letter P (for partnership) while those driven by attitude 2 are shown by the letter D (for dominant) in the below panels. These cultures can be thought of as individual tribes struggling to survive with respect to their local resource base as denoted by the green circles in Fig. 5.

The evolutionary sequence for Panel A consists of the following steps:

1. Cultures P and D are randomly placed such that the distance between them precludes either physical interaction or mutual knowledge of their existence. However, members of each tribe in their attempts to make a map of their world can likely walk around in any random direction and not run into a boundary. Thus very early on, the perception of the world as **infinite** can form.
2. Around each culture, a circle is drawn and the radius of that circle defines the sustainability horizon, or the geographic area over which there are sufficient resources to support the needs (discussed in section I) of the tribe for many generations.
3. Culture P is able to "sustainably develop" within their radius and therefore have no need to harvest resources beyond their radius (of course, they may wish to explore beyond their radius) as long as they adhere to their partnership principle. In general, culture P will live in equilibrium with respect to their available resources.

   Not all culture P's will survive in their initial location as some kind of environmental catastrophe (extended drought, flood, earthquake, volcano, etc.) may destroy most of the resources in their sustainability radius. On the Earth, these kinds of events are more common than people realize and include:

   - The dispersal of local culture by the Missoula Floods of 13000–15000 years ago;
   - The Black Sea deluge event of 7500 years ago (see Ryan and Pittman 1999) in which glacial meltwater substantially raised the water level in the Mediterranean Sea relative to the Black Sea. This hydraulic head pressure then collapsed the land barrier between these two seas and opened up the straits of Bosporus. Salty water from the Mediterranean then poured into the black sea which rose about 6 inches per day for 2–3 years thus dispersing any culture that may have been located there.
   - The collapse via drought of the Akkadian Empire of Mesopotamia about 4200 years ago;
   - The impact on and dispersal of the Minoan Culture due to the 1651 BC eruption of the Santorini volcano located near Crete;
   - The collapse of the Mycenaean (early Greek) world around 1100 BC likely due to a series of large earthquakes (see Nur 1997). The Mycenaean cultural centers of Midea and Tiryns were built on the top of cone-shaped limestone hills that could amplify seismic waves. Archeological evidence shows that some of the cities in the Mycenaean world were destroyed by fire during this time (and large earthquakes can easily knock over assorted oil lamps, etc., and start large fires).

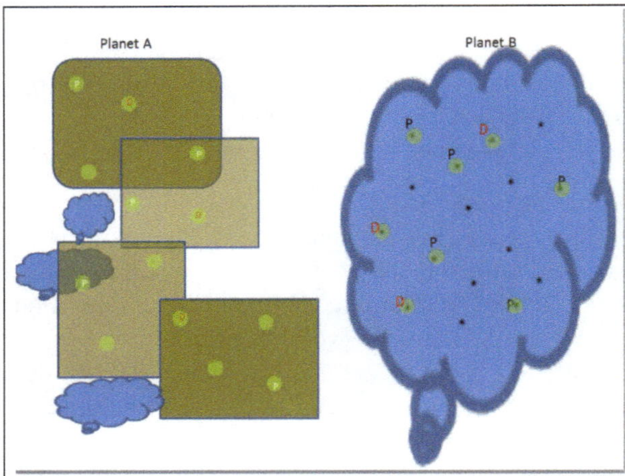

Initial Starting conditions where 5 tribes P and 3 tribes D are randomly placed on the respective world maps. The green circles donate the size of the sustainability region (radius, as described in the text). No individual tribes knows of the existence of others.

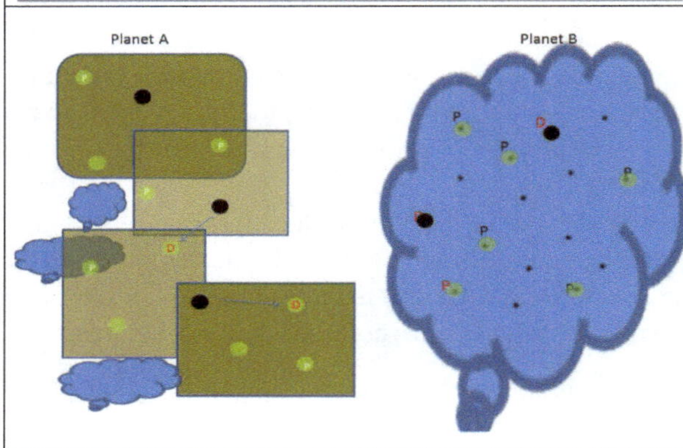

As described in the text – at some time later sustainability radii of the D tribes have been used up. In the case of the continental world, 1 D tribe has gone extinct and 2 have randomly migrated. In the island world, 2 D tribes have gone extinct and 1 has converted to P.

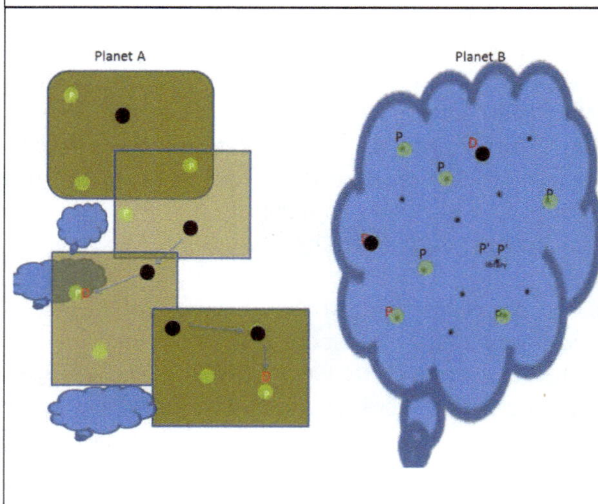

In this final stage, cultural mixing is now occurring. In the continental world, the random migration of D brings it into contact with P (who hasn't moved). In the Island world there is a random exploratory contact between scouts of the respective P culture – they meet on a random island and build a library (see text).

Fig. 5 Evolution to different cultures (P and D) with respect to their use of resources on two different planetary surfaces: Plant A is comprised to large continental land masses separated by volumes of water while Planet B consists of nothing but small islands distributed in a vast ocean

- The collapse of the Mayan culture around 900AD due to an extended 150 year dry period with four distinct drought phases, each lasting for 15–25 years.

Hence there are drivers that cause the P locations to change and this might cause cultures P and D to mix at later times (see more below).

4. The likely scenario for culture D is to relatively rapidly deplete the resources contained within the sustainability radius. This will then motivate culture D to move to another area and begin to exploit that area. The key here is that resource availability to culture D is large and relatively easy because the connected continent resource area is thousands of times larger than the area in an individual sustainability radius. Hence, in this initial sparsely population of tribes, it is easy for culture D tribes to become nomadic resource hogs.
5. Note also that we consider it likely that, in the desire to harvest resources more efficiently, culture D will eventually have an industrial revolution. In the case of culture P is seems less automatic that they will engage in an industrial revolution since they have already established sustainability for many generations and therefore do not have the need to harvest resources at an increasing rate.

Given the nomadic nature of Culture D, at some point they will randomly encounter either another D tribe or a P-tribe. These encounters are shown in Panel c for Fig. 5. When this happens, that particular tribe is no longer isolated but has mixed with another culture. There are different possible outcomes of this mixing encounter which have an ultimate effect on resource harvesting of planetary resources.

- The D-D mix: In this case, the mixing is most likely occurring because of two expanding cultures. On this planet, that has often resulted in warfare between two dominant cultures. The end result of such warfare is completely counter-productive—the cultural values of the D-tribes have no yet changed and their desire for better weapons may accelerate the discovery and use of technology. It does absolutely no good for one tribe D to steal the resources from the other tribe D since they are destined to use those acquired resources up pretty fast. The end result (see Panel d) of this process is planet wide occupation by some tribe D that now has a sustainability radius the size of their entire planet. Now what?
- The P-P mix: This would be a rare event and would only because by some environmental catastrophe that displaces a tribe P. Statistically it is far more likely that tribe P runs in to a tribe D long before it finds another tribe P. However, on this planet of large continents, if two tribe P's mix they are likely simply to wave and say hello while one of the tribes stays in its sustainability radius and the other tribe establishes a new area. If not too widely separate, perhaps the two tribes will periodically visit each other and exchange ideas. Note that the P-P mix will have a far more interesting consequence in the island world.

- The D-P mix: Here there will be a range of interesting outcomes ranging from cultural extinction to the formation of a new culture. Here are the possibilities:
    1. The extinction of culture P—culture D simply shows up one day determined to steal the resources from culture P and to possible annihilate culture P because culture D carries bigger sticks or has developed technological weaponry. On Earth, the expansion of the Mongo empire in the time period 1200–1300 AD from a relatively isolated region to all the way west to the Mediterranean sea and East to the entire Chinese coast provides a real example of cultural expansion (largely through the use of horses and warrior horseman as a technology advance). In this case, the planet now has less examples of culture P and larger examples of culture D.
    2. The extinction of culture D—while this is very unlikely it is not entirely impossible as cultural mixing can also produce large scale disease propagation. Indeed, this mechanism is likely the only available to culture P to preserve itself. Of course, culture D may develop anti-biotics and thus revisit culture P at some point in the future.
    3. The emergence of the culture PD—this is the most interesting outcome and requires that the values of culture D change because of the values inherent in culture P. That is, culture P has served to educate culture D, so that culture D decides to avoid the inevitable Mongo empire problem and the need for continual expansion. In terms of the theme of this chapter, perhaps the cultural mixing has forced culture D to learn and practice *sustainable development*.

Now let's consider a similar kind of evolutionary scenario that occurs on the island planet as shown in Panel B:

1. Cultures P and D are randomly placed on individual islands such that the distance between them precludes either physical interaction or mutual knowledge of their existence. However, strongly unlike the previous case, individual tribe members can easily discover a boundary of their world and rapidly understand that their world is quite **finite**.
2. For each culture the sustainability radius is equal to the size of their island.
3. Since Culture P is able to "sustainably develop" within their radius and therefore have no need to harvest resources beyond their radius then they can manage living on their island without resource conflict. This is in stark contrast to culture D who is destined to rapidly use up their island resources and become extinct. For example, culture D will likely harvest all the timber resources on their island before they realize that they need to build a boat!
4. In this scenario, Culture D becomes immediately challenged by their environment. If they cannot quickly become sustainable they will be extinct. Forced sustainability then provides a path way for a quick transformation from Culture D values to Culture P values. Hence, this planetary surface has a natural mechanism to prohibit the appearance of the Mongol empire. Perhaps this is one of the reasons why there never has been any Polynesian empire.

5. Eventually an island P tribe begins to explore beyond their boundaries due perhaps to innate curiosity or maybe waste from an extinct culture D has washed up on their shoreline. In either event, culture P still has timber resources (as they have not depleted them) so one day they build a boat and sail in some random direction.

The form of random cultural mixing on this planet is now very different than before. Since all the D cultures have become extinct of have been transformed to P cultures in order to survive, we only will have a PP culture mix. Since it seems very unlikely that either P culture will be carrying big sticks, then the initial PP encounter should be motivated entirely by curiosity and idea exchange. In all probability the two P cultures will then learn and grow from each other. Perhaps they will discover an uninhabited island and build a library there for all future P cultures to come and visits and make contributions. The main point of this hypothetical example should be obvious. If resources are known to be finite at an early stage of tribal evolution, this is likely to produce better behavior and more sustainable practices than in the case where a seemingly infinite resource base is laid out for your exploitation. The reality of the Earth as an island that has to sustainably support all its inhabitants, became fully clear on Dec 24, 1968 with the first Earthrise image obtained by the crew of Apollo 8 as, for the first time, the Earth is now seen from an external vantage point. This is a view of the world that Culture P has always known to be correct (Fig. 6).

We can further distinguish culture D and culture P values in the context of planetary subjugation and climate change We argue that our continued reliance on mined resources as the principle source of electricity and energy generation carries with it a tremendous environmental cost as well articulating a consistently

Fig. 6 Earthrise from Apollo 8, 1968

inappropriate message that we exist to dominate the planet. While there are important engineering differences between electricity generated by wind compared to natural gas fired electricity, such as overall efficiency, capacity and reliability there is, perhaps, a larger symbolic difference. When we mine resources (dig precious things from the ground) that is largely an action which is invisible to the end user—most all of the infrastructure is below the ground, the pollution from fossil fuels is mostly invisible CO_2, and on the individual human scale, the Earth is essentially infinite so what is the harm of digging a hole or two? This is fully consistent with the evolution of Culture D as previously described since metaphorically; digging up resources represents dominance on the part of the digger.

In stark contrast is the potential harmony of a spinning wind turbine. In this case, we only get electricity when the wind is blowing and therefore are in a necessary partnership with nature for the use of that resource (e.g. culture P). Such a partnership tends to foster better resource management and resource equilibrium. Using machines to dig holes in the ground from a seemingly infinite resource, unbinds us from any notion of partnership and allows resource consumption to steadily increase. Commensurate with this increase is the previously described increasing rate of pollution (greenhouse gases) which is accelerating the rate of excess build up in the atmosphere resulting in climate change and increasing climate volatility.

Despite these considerations, fossil fuel production out of drilled well-heads over the last 10 years in the US continues to escalate as shown in Fig. 7. This escalation is primarily the result of tapping new, short term sources of oil. These are described in the figure below as "tight oil" (e.g. North Texas oil fields, Bakken formation in North Dakota) and the increasing temporary available of natural gas through the hydraulic fracking process. This figure clearly demonstrates the condition of business as usual (BAU) land is strongly consistent with the value system of culture D. In the real world, despite all the rhetoric about sustainability we continue to arduously pursue new and unconventional sources of oil which are difficult to

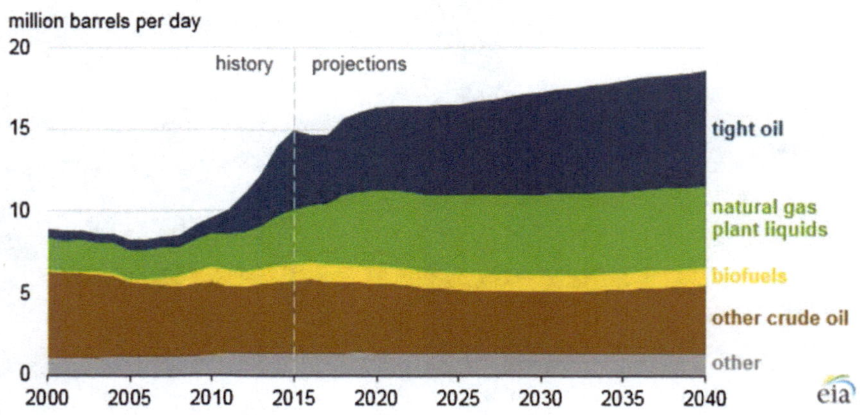

Fig. 7 U.S. production of petroleum and other liquids (2000–2040)

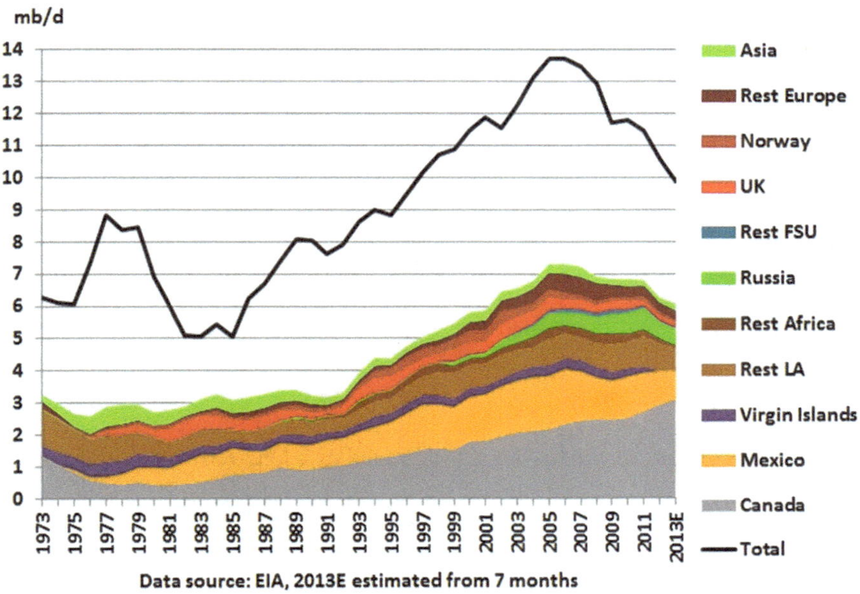

Fig. 8 U.S. crude product imports from Non-OPEC

extract and during the extraction process do considerable local environmental damage. The fuels extracted from this BAU approach, of course, are burned to further accelerate climate change. All of this seems quite unwise and yet the momentum cannot be stopped.

We have been in a position like this before regarding choosing whether to continue with BAU or to invest in new and more renewable forms of energy. Figure 8 shows the crude oil import profile of the US over the 40-year period 1973–2013. For reference, current use of oil in units of millions of barrels of oil per day (mb/d) is about 21; hence in the mid-2000s US imports were 60–65% of daily use. We start with the early 1970 s because that is when North American production of conventional crude oil peaked (e.g. the Hubbert Curve—see http://www.albartlett.org/articles/art2000jan.html). Over the period 1975–1990 US demand for oil fluctuated in the range of 16–18 mb/d. Since domestic supplies were limited, then by 1975 about 50% of our oil use was imported exclusively from the Persian Gulf region. Congress expressed concern about energy independence and posed this heavy reliance on imported oil as a security risk. Hence, drilling and exploiting crude oil resources in Alaska was justified and that was chosen over efforts to reduce our dependency on crude oil through better fuel efficiency as well as the use of renewable energy resources. As discussed above regarding the perception of the

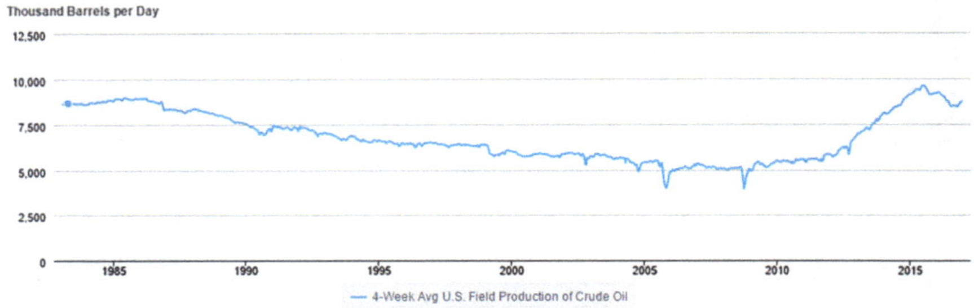

Fig. 9 Data from the Energy Information Administration (EIA) showing the 4 week average US production of crude oil from 1980 to Sept. 2016 from various US oil fields—see also http://www.eia.gov/petroleum/imports/companylevel

seemingly infinite resource, Alaska was regarded as a long-term solution to our oil importation issues. But this was not a long-term solution and by 1990 we had return to the levels of imported oil required in the mid-1970s. Thus, harvesting Alaska provided the US only a limited, 15-year window, of increased domestic supplies of crude oil. After that window was closed, we were right back to square one as we have required the BAU trajectory. Well, as of 2012, the US is doing something very similar to the exploitation of Alaska oil resources. Now, we are choosing to harvest short term unconventional crude oil in the US so as to once again decrease our dependence on foreign oil instead of investing in renewable energy sources at a scale which is comparable to our use of crude oil as an energy source.

In Fig. 8, the trajectory from 2009 through 2013 is favorable. Continuing that trajectory, however, would require that we are only in the initial phases of extracting oil from very large unconventional resources. Alas, these resources are not very large (with respect to our demand) and hence accessing them is just another short term "fix" while we maintain our addiction to BAU. Figure 9 provides the most recent information on US field production of crude oil and shows that this production peaked in July of 2015 at 9.6 mb/d and since that time has fallen by about 1 mb/d to the current level of 8.6 mb/d. Thus, imports of crude oil to the US are once again rising as, like Alaska, these new sources of "infinite" supply, are in fact quite limited.

Overall, we stay on the BAU trajectory through continual drilling of new well head to tap unconventional sources of fossil fuels such as shale oil and shale gas (via fracking). This is certainly no partnership with nature as our dominance continues. In the larger view of public trust, harmony with nature would seem to be mandatory yet we dig and dominate and disregard instructions from previous and ancient generations:

> When we dig precious things from the ground, we invite disaster—Hopi Prophecy

4 A Historical Consideration

The fundamental question posed by the issue of sustainability or sustainable development is this: **Are we in a partnership with nature?** Again, prior to the industrial revolution this question is mostly moot because partnership was generally forced on most human cultures. Once we develop the steam engine, we now have the ability to overcome some of the boundaries set by nature. In this process of subjugating nature do we ever consider that harm to nature is harm to us? Descartes (1635) considered the role of nature in his treatise *Discourse on Method.* Descartes's view is that the Universe is a clockwork machine and by methods of reduction the individual observer can discern the workings of the clock. In this view, Descartes treats nature as a **machine** that simple responds to the orders of the clock workings. Nature herself is not living and has no soul and therefore it is impossible for the civilized man to have any kind of spiritual relationship with the Earth. Descartes view is expressed through many of his writings and quotes in which he directly declares that nature/animals have no intelligence, that nature is a lifeless machine, and that humans (man) are entitled to preside over nature:

- It is also a very remarkable fact that although many animals show more skill than we do in some of their actions, yet the same animals show none at all in many others; so what they do better does not prove that they have any intelligence, for if it did then they would have more intelligence than any of us and would excel us in everything. It proves rather that they have no intelligence at all, and that it is nature which acts in them according to the disposition of their organs. In the same way a clock, consisting only of wheels and springs, can count the hours and measure time more accurately than we can with all our wisdom.

  I suppose the body to be just a statute or a machine made of earth
  And thereby make ourselves, as it were, the lords and masters of nature.

The result of Descartes' work, together with others like Newton, was to produce what has been called the mechanical philosophy. This philosophy makes nature nothing but a machine—any purpose or spiritual significance to nature did not exist. In the mechanical philosophy nature is viewed as merely lifeless matter in motion. This idea was later strongly enhanced in the publication of James Hutton's 1788 publication, Theory *of the Earth* in which the Earth is generally described as a machine, albeit a complex one, whose purpose is to produce a habitable world and sustain life.

Thus, at the time of the industrial revolution, the prevailing European world view, is that nature provides a resource to man who is entitled to harvest it in any way that seems fit. The use of machines as a harvesting agent seems fully justified in this world view and the moral issue of damage to the Earth, its other species, and its resource base is not considered. Of course, the eventual scale of the machine domination of the planet would certainly be beyond anyone's limits of comprehension in the 19th century. Nonetheless, the philosophical and operational roots of our BAU trajectory are firmly set at this time.

Indeed, in his 1865 essay on *The Coal Question*, Stanley Jevons makes the following points:

- *It is wholly a confusion of ideas to suppose that the economical use of fuel is equivalent to diminished consumption.* The very contrary is the truth … Every improvement of the (steam) engine, when affected, does but accelerate anew the consumption of coal.
- *Coal in truth stands not beside but entirely above all other commodities. It is the material energy of the country-the universal aid-the factor in everything we do. With coal almost any feat is possible or easy; without it we are thrown back into the laborious poverty of early times.*
- *For once it would seem as if in fuel, as the source of universal power, we had found an unlimited means of multiplying our command over nature.*

The relatively large and accessible coal deposits of the UK would allow them to rapidly increase their world dominance and economic development. The waveform (Fig. 10) of coal exports from the UK easily shows that when a new resource is found, it is developed as fast as possible rather than to consider a more sustainable approach to harvesting the resource. The data show that total exports of coal increased by approximately a factor of 10 over the period 1820–1900, a completely unstainable rate of production that will eventually exhaust the coal resources of the UK to lead to a decline in exports. Overall this wave form of rapid growth is similar to what we showed earlier with respect to container traffic and once again is a sign of unsustainable practices.

In the US, coal consumption also exhibited strong escalating growth over the period 1850–1920 as shown in Fig. 11:

The period of most rapid growth occurs from 1875 to 1920 as coal use increased by a factor of 10 over this 45-year time scale—a clearly unsustainable rate. Over this time period, the growth rate is also fairly continuous with only a small alteration seen at the time of World War I. While the US did build a robust manufacturing infrastructure, primarily in cities with access to ports (Great Lakes, East Coast) during this time, by 1930 the limits of coal dominance became apparent as (a) most cities in the eastern US had extremely levels of coal burning pollution

Fig. 10 UK Coal exports 1800–1950

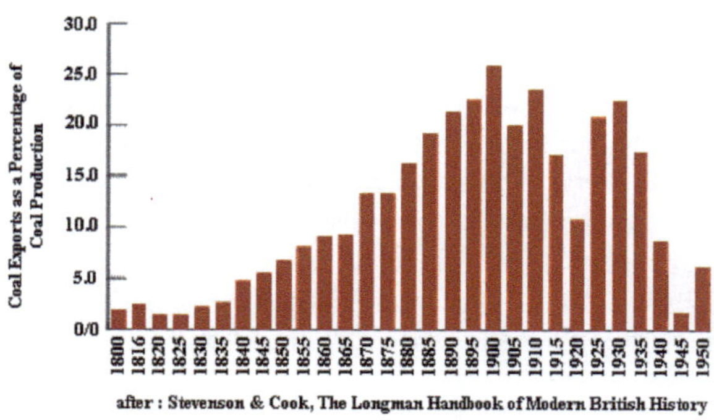

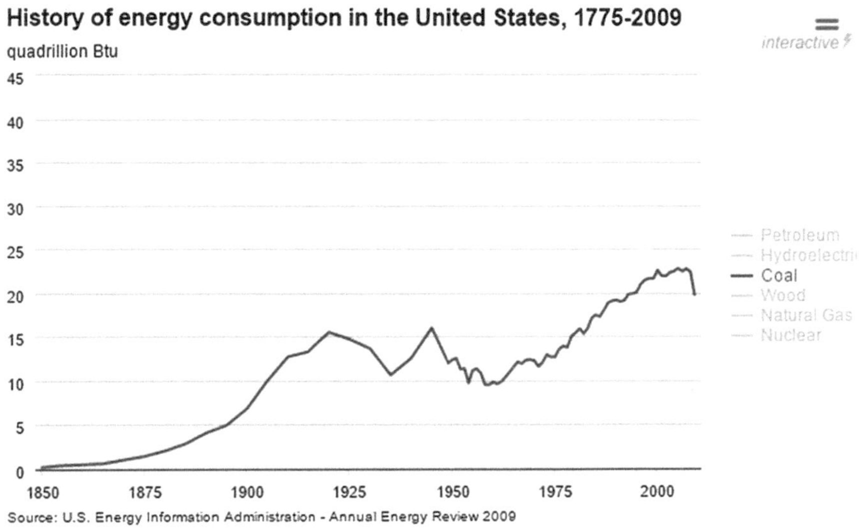

Fig. 11 The history of coal consumption (only) over the period 1850 to 2009—here we are only interested in the period 1875–1920

which posed serious health risks (example on the left comes from Pittsburgh in 1940); and (b) the US was entering an extended period of economic depression following the catastrophe market crash of 1929.

Enter now the ideas of Lewis Mumford a prominent social critic (especially of architecture) who in the publications entitled "Technics and Civilization" and "The Culture of Cities" makes a set of arguments about the use of hydropower as a cleaner source of electricity. Mumford argues that the current era (which he calls the Palotechnic Era) relies upon "coal and iron" and exploited finite fossil fuels that generated a "befouled and disorderly environment; the end product an exhausted one". He strongly promoted the potential of hydro-electric power as an "inexhaustible" supply which would re-open the Western frontier and rekindle the frontier experience that "makes this country great". Mumford wrote about the role of technology and the specific role of electricity in shaping society. He felt that hydro-electric power was liberating and he called this era the Neotechnic Era where this "infinite" supply of hydro power would purify the social and ecological environment. Furthermore, that hydro power would be developed by the Federal government as public power because that would further promote democracy. He envisioned an American West in which growth would occur around regional river basin development using dams as the principle means of development. He argued that such a power source would produce a cleaner and more "pastoral" living style that would liberate citizens from the de-humanizing sate of over-industrialized and over-crowded cities in the East fed by the continual burning of coal and iron. In essence, Mumford made an argument 80 year before the present argument made in the sustainability movement: fossil fuels (e.g. the Palotechnic) are evil and green energy is a more democratic way to provide Americans their required power. Through the Federal Dams building projects of the 1930 s (which employed hundreds of thousands of workers during the Great Depression) America was becoming greener. By 1940 an astounding 35% of the total electricity came from the renewable source of hydropower and 75% of the electricity available to the American West was hydro based.

Well, whatever course we were on during the 1930s got abruptly changed due to World War 2 and the overall effect of the post WW2 world on consumer consumption is generally not well understood as being the principle driver for rapid resource depletion and climate change. Figure 12 is taken from Bohanon (2009) (https://www.mercatus.org/system/files/PostWWII_Recovery_Bohanon_MOP112-%281%29-copy.pdf) and shows the very rapid onset of private dollars invested and used in consumption, as it compares the years 1944–1947.

The reason behind this explosive growth has to do with the gigantic production machine that the US built in order to build mass amounts of munitions to win the World War 2 conventional ground war. US production was at an all-time high (see Fig. 13) and when the war was over, there was no desire to turn this production machine off. In the words of Robert Nathan, a white house economist in 1944:

> Only if we have large demands can we expect large production. Therefore, it is important that in planning for the postwar period, we give adequate consideration to the need for ever-increasing consumption on the part of our people as one of the prime requisites for prosperity.

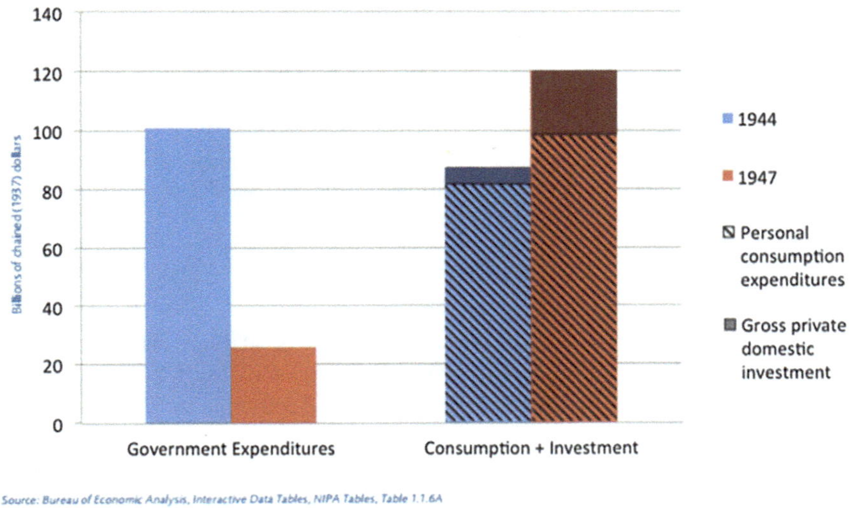

Fig. 12 Showing sharp consumption rise from 1944 to 1947

Fig. 13 The rise of manufacturing arms for WW 2

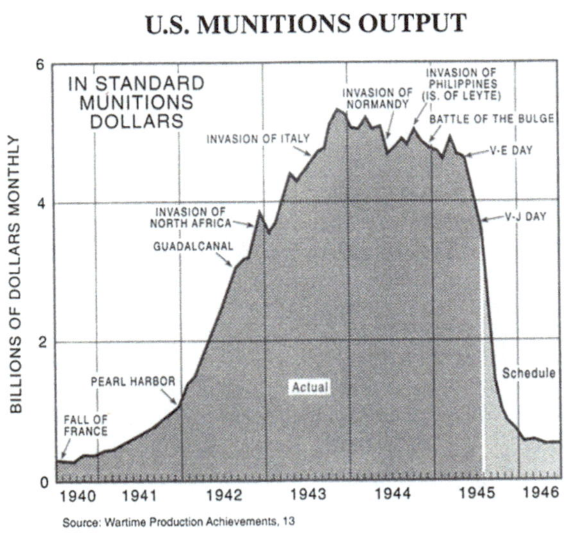

Note the mandate of "ever-increasing consumption" as that which defines the American lifestyle. This is about as far away from sustainability as one can get. In order to motivate this consumption, there was a gigantic advertising campaign (particularly on the newly developed technology of television) that Americans become happy when they are consuming (this is well detailed in the book called

Advertising in the Age of Persuasion: Building Brand America 1941–1961—D. Springer 2011—see also the treatment of Cohen 2004). The quote below is a particularly good summation of this period and the legacy of consumption which follows:

> Our enormously productive economy demands that we make consumption our way of life, that we convert the buying and use of goods into rituals, that we seek our spiritual satisfactions, our ego satisfactions, in consumption. The measure of social status, of social acceptance, of prestige, is now to be found in our consumptive patterns. The very meaning and significance of our lives is today expressed in consumptive terms (Victor Lebow 1955 —The Journal of Retailing)

The end result of this process is consumer non-linear scaling. That is, the use of some product does not scale linearly with the number of users but scales as $n^{P;}$ $P > 1$; A good example of this is to look at US electricity use over the period 1950–2000:

US population	US electricity consumption
1950: 151 Million	1950: 400 TWHs
2000: 281 Million	2000: 3500 TWHs
Growth = 1.86	Growth = 8.75

In this case, n = 1.86 and P is the value that satisfies $(1.86)^{P} = 8.75$; or P = 3.5. Hence, under this scaling if population were to double (n = 2) the demand for electricity would increases by $n^{3.5} - 11$. This is strongly non-linear growth and most products scale in a similar non-linear manner with respect to the growth of the number of consumers. This, of course, accelerates resource depletion and also indicates that growth is occurring primarily through increases in per capita consumption as clearly shown in Fig. 14b over the period 1960–2000 for various regions in the US and the US as a whole (green line). Although per capita consumption now (2017) in the US appears to be flat, we did go through this explosive growth stage in electricity use primarily fueled by production and consumption. One final form that shows explosive non-linear growth is in consumption of bottled water. Over the period of 1976 of 2012 (note the small decline induced by the economic meltdown of 2009), consumption increased by a factor of about 15 while population increased by a factor of 1.44 (314/218 million); in this case P – 7.5. When discussing sustainability of any resource it is paramount to use data to discover the rate at which this resource is consumed; only then can you come up with a kind of public policy that would quantitatively curb the consumption of that product. Ultimately this unbridled consumption is putting sufficient energy into the Earth systems such that climate change will be a natural response. We discuss this response briefly in Sect. 6.

(a)

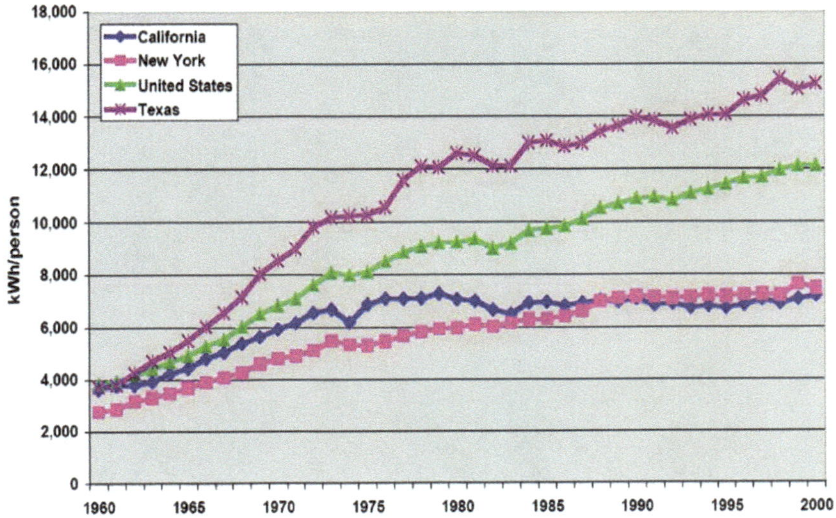

(b)

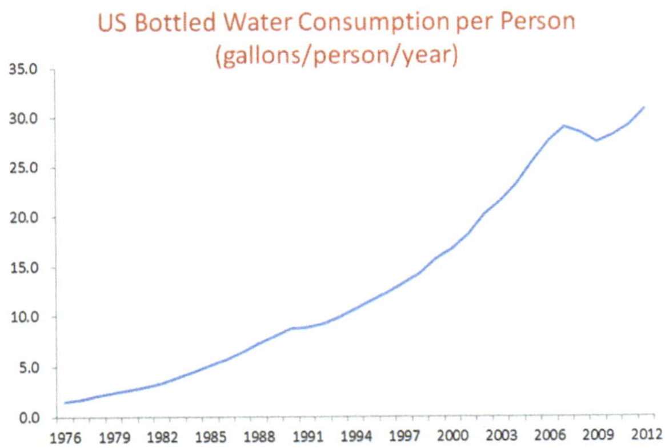

Fig. 14 **a** Growth in per capita electricity consumption over the period 1960–2000 in the US total and in California, Texas and New York. **b** Growth of bottle water consumed 1976–2012

5 What to Do: Achieving a "Just Sustainability"?

While there are many challenges to achieving sustainable, two seem foremost: (a) as argued, sustainability is a quantitative problem and effective policy would be aimed at lowering resource consumption to the point where it is quantifiably impactful (in a positive manner) on the Earth system. Unfortunately, this is rarely

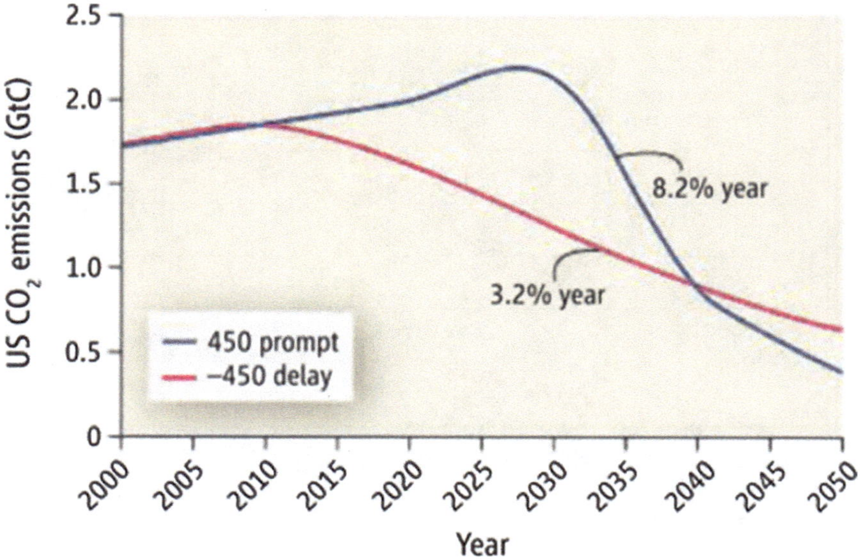

Note: The key is mislabeled in the graphic. The red curve is prompt and blue is delay

Fig. 15 CO_2 emission decline rates needed to meet the 450 ppm stabilization level by the year 2050: Source is the 2011 Whitehouse climate report

discussed and never done. One potential example of this quantitative approach is in the area of CO_2 stabilization in which real numerical emission targets, in units of tons of carbon emitted per year, must be enforced in order to reduce atmospheric CO_2 (from 450 ppm) over some timescale. However, in the real world we have paid no attention to this and now we are at the point that where CO_2 emissions need to experience dramatic and continuous reductions (see Fig. 15). Since we have no yet started any reduction, we are very much on the blue curve of continuing emissions which then will require a drastic reduction to reach the 450 PPM stabilization target and this is very unlikely to occur; (b) there is a strong need to make issues of equity and social justice more prevalent than the issue of climate change. Overall, we still live in mechanical philosophy value system where we cannot conceive that nature has any value beyond being a resource. This must change. Sustainability is not about the more efficient harvesting of resources, its about establish a more equilibrium use of resources with respect to the innate planetary cycles. This perhaps best articulated by Agyeman et al. 2002:

> Sustainability cannot be simply a 'green', or 'environmental' concern, important though 'environmental' aspects of sustainability are. A truly sustainable society is one where wider questions of social needs and welfare, and economic opportunity are integrally related to environmental limits imposed by supporting ecosystems.

This approach has led to the concept of achieving a 'just' sustainability which is summarized in Agyeman et al. 2003:

> The need to ensure a better quality of life for all, now and into the future, in a just and equitable manner, whilst living within the limits of supporting ecosystems.

Perhaps the most important feature of this concept lies in the recognition of environmental limits and taking care not to exceed those limits that put ecosystems and therefore other species in danger. In view of this, we can define the components of this just sustainability as follows:

1. Improving our quality of life and well-being:

In a strictly economic system quality of life is generally concerned with standard of living only and does not address notions of social justice, dignity and equity. I suppose that one assumes those three items are a natural consequence of economic development which yields increasing prosperity. In turn, this increasing prosperity is generally measured in a single indicator: GDP. This then implies that before the existence of GDP people had to generally be unhappy. This of course is absurd (see one of the student exercises that elaborate on this issue). Two things need to be developed to improve upon this economic measure of quality of life: (a) a more holistic view of prosperity needs to be adopted so that consumers become aware that continued economic growth is not a requirement for increasing prosperity; (b) progress needs to be measured in something other than GDP and should include issues like environmental and public health as well as equity and social justice.

2. Being able to continually meet the needs of future generations.

This echoes the basic Brundtland report of 1987 but in view of point 1 above can now be expanded to include justice and equity as needs. This is also a quantitative issue in terms of needs per capita and material consumption. One could even imagine setting some kind of consumption tax, when individual consumers exceed their baseline allocation. The 1950 s certainly set the stage for human social identity being determined by consumption and all the advertising agencies cheered us on, to the point where something like Tupperware suddenly became a fundamental need (see http://www.pbs.org/wgbh/americanexperience/features/general-article/tupperware-consumer/). This process of continually defining social needs by consumption leads directly to a kind of social inequality that ultimately does significant damage to various social structures (see Wilkinson and Pickett 2009). One of the frameworks for a just sustainability is to replace consumption based social identity with something more meaningful and real.

3. Increasing recognition that justice and equity are hallmarks of an advanced civilization.

The concept of justice is complex and nuanced (Sen 2009). Schlosberg (1999) was one of the first to attempt to map these concepts onto the idea of environmental justice. The meaning of environmental justice (e.g. Shiva 2015) is now widely discussed and has now expanded to the domains of climate justice and ecological debt (Schlosberg and Collins 2014; Warlenius et al. 2015). While this issue maybe complex and nuanced, it does seem to distill to one reality—the Earth's climate and energy resources should be equally available to all its citizens if we truly have a just global society. Data clearly shows this is quite far from the case and differential consumption of resources, by individuals and countries, leads to the kind of enormous inequity discussed in Sect. 1. As long as the global playing field remains in this extremely unleveled situation, progress on important environmental issues is likely to be strongly impeded.

4. Living within ecosystem constrains and limits.

Figure 16 shows the latest version of the Global Ecological Footprint as measured over the period 1960–2012 as provided the 2016 Living Planet Report (http://awsassets.panda.org/downloads/lpr_living_planet_report_2016.pdf). The meaning of this diagram is that the Y-axis contains the annual earth resources needed for one years' worth of consumption. The dotted green line is the (slowly increasing) availability of Earth resources to support annual consumption. If the total consumption were less than the green line, then we would be consuming within our ecosystem limits. Often times, for greater impact, the Y-axis of this diagram is in

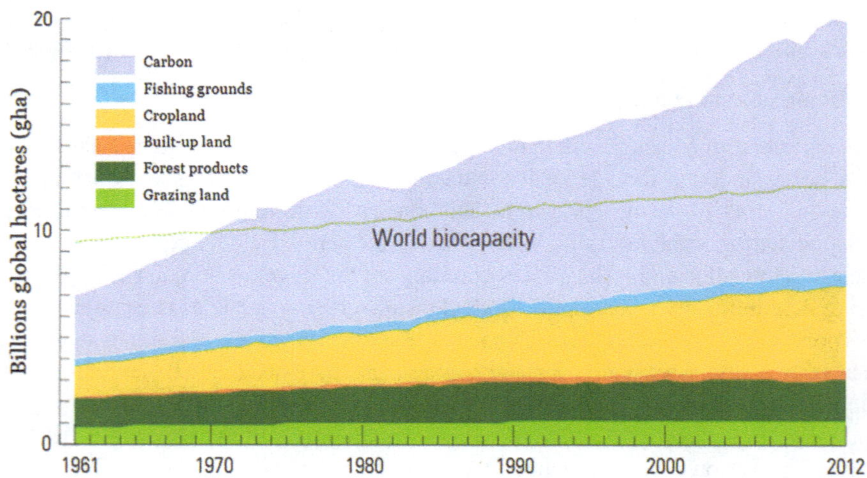

Fig. 16 The ecological footprint of humanity is shown in the most recent Living Planet Report

units of Earth masses. **In 2012, 1.6 Earth masses were needed to provide the resource base for global consumption in that year**. However, ever since the early 1970 s we consume more in a given year than the Earth can provide. Because of the presence of temporary buffers in the system we can engage in this overshoot behavior for some time. For example, we can mine the ocean for decades, in the sense that we harvest more fish than the oceans can replenish. But we have been engaged in this overshoot for 40 years by now and consequences in terms of habitat and species loss are becoming noticeable (e.g. Tittensor et al. 2014), The most direct manifestation of our overshoot action is to displace Carbon in its natural earth reservoirs and relocate that overshoot excess into the atmosphere where the accumulation is increasing. Of particular note in this Figure is the change in slope of the grey envelope occurring around the year 2003—this is when the Chinese economy being to emerge as a global consumer as well as when greenhouse gas emissions from China began to exceed the US. The slope continues unabated through 2012 despite the presence of the global economic meltdown of 2009. The small decreases in particular years shown below also mark short term recessions (e.g. 1980–82 event). Overall, however, we have made zero global policy progress on reversing our global footprint, a concept first presented to us in the 1972 Club of Rome Report (Meadows et al. 1972).

Our inability to consume within our global resource limitations could be related to the culture D versus P story discussed above. We simply do not acknowledge that our resources are finite and operate under the implicit assumption that new technology will open up new resources. Indeed, there is some truth to that as new deposits of unconventional oil have been discovered as a result of improved geophysical imaging techniques. And so we exploit those, as mentioned above to maintain short term BAU, and thus accelerate the overall impact on climate change. This is clearly beyond stupid but this is the same civilization where there is talk about eventually mining the asteroid belt for precious metals. This very much is in line with inevitable result of culture D—they have mined their sustainability radius and now are looking for an even bigger one—when does that process end?

The operational idea of ecosystem limits has been well discussed by Rockström et al. (2009) which is essentially a modern version of the 1972 Club of Rome report. From this work a vicious cycle seems to be in play. As resource availability becomes an increasingly fundamental limit to economic growth, individual countries then attack the problem with increasing severity. In the US, blowing the tops off of Appalachian Mountains to mine coal would seem to be an act of desperation. Similarly, drilling for oil under that is located under 5000 feet of water and 15,000 feet of marine sediments in the Gulf of Mexico also seems extreme. If you really need a 20,000 foot long pipe drilled into the ground so that you drive around and contribute further to climate change, something is fundamentally wrong!

6 Climate Change: The System Response to Unsustainable Consumption

The accumulation of atmospheric CO_2 is the direct manifestation of our strongly non-equilibrium consumption of planetary resources. However, there is another measure that is even more fundamental and scientific to the root cause of climate change. Ever since the industrial revolution human activities have produced thermodynamic waste heat. Burning fossil fuels is just one form of producing industrial waste heat. Figure 17 shows how waste heat gets redistributed throughout the various planetary reservoirs. Virtually all waste heat ends up in the oceans. For instance, if you were to light a candle that excess heat that you added to the air in the room you were in, would eventually get carried to the oceans as precipitation takes that heat out of the air, deposits it in the soil, where it eventually runs off to the oceans. The timescale for this process can be very long (decades) because heat can be stored in various buffers (e.g. continents) in the Earth system but eventually that all ends up in the world's oceans.

Since 1957 we have had an ocean buoy network to measure sea surface temperatures. Scientists have now been able to transform that temperature date into ocean heat content. This is a difficult calibration to make and different groups use different procedures but they all end up showing that ocean heat content is now systematically increasing. This is shown in Fig. 18. We are now systematically dumping more heat into the oceans that it can dissipate via its natural channels. This

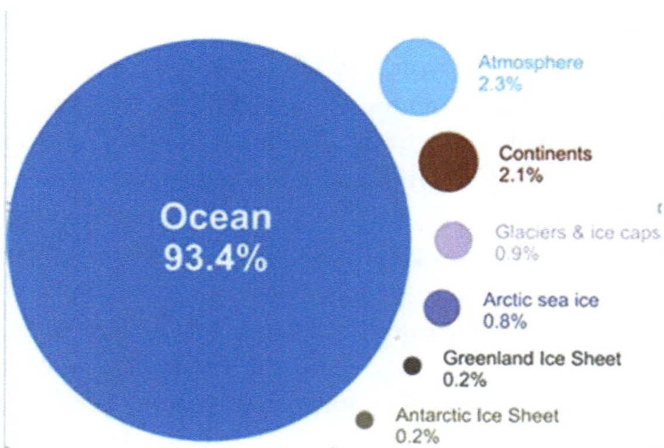

Fig. 17 The distribution of waste heat among different systems of the Earth

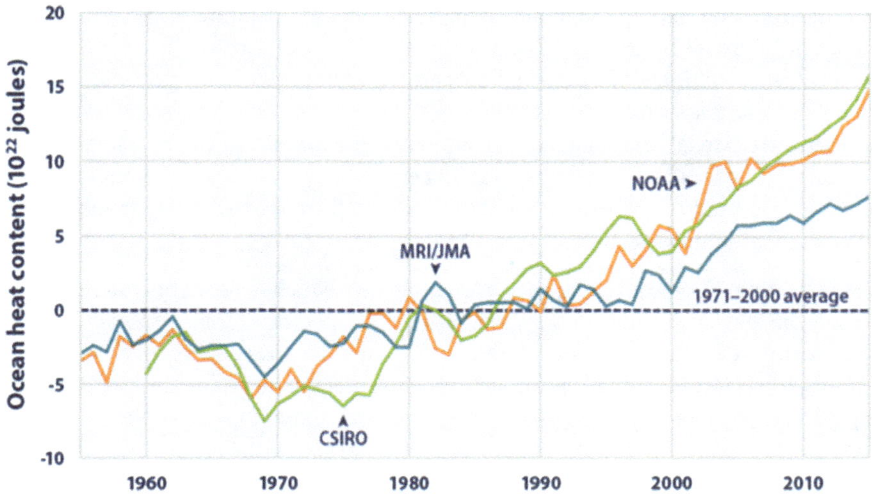

Fig. 18 Measure of heat content increase in the world's oceans by different agencies. Details of this figure are described in the text

excess heat couples directly to the world's climate system and adds significant energy to it resulting in not only climate change but increasing weather volatility. It is the heat content of the oceans and the movement of that heat content which is the principal driver of the earth's climate system. If you perturb that system then you will perturb (perhaps strongly) the overall climate system of the Earth.

Key components of this figure are:

- The dashed line represents the 1971–2000 average value. Much like the overshoot phenomena discussed previously, ever since about 1990 industrial activities have added more waste heat to the oceans than the system can handle. As a result, the signature of climate change driven by excess ocean heat would start to emerge around 1990.
- The response of the oceans to this exist heat are to (a) transport some of it vertically down thus changing the vertical temperature profile of the oceans, (b) change its surface heat redistribution via ocean current amplification, (c) directly transport it to atmospheric heating.
- The oceans are 850 times denser than the atmosphere and thus can hold an enormous amount of heat. Indeed, it is this high heat capacity of the oceans that has served as a buffer against increasing warming on the land—that buffer is now saturated. If the entire excess heat content of the oceans could be

transferred directly to the atmosphere, the global temperature of the atmosphere would increase by 36 C! It is the large buffer capacity of the oceans that very much help mitigate the potential catastrophe that is posed by periodic ice ages on the Earth's surface.

- Increased ocean heat content will ultimately be carried to the poles and accelerate the loss of sea ice as well as lowering the permafrost thickness of most soils (especially in Siberian). As shown below, there is now strong evidence that this affect is now occurring.

On the human timescale, our ocean heating rate, which is now equivalent to each person the planet burning one gallon of gasoline per day (somehow we seem to deserve this equivalency), means that we have now irreversibly overheated the oceans. Even if we turn off all industrial waste heat channels today, we will still experience decades of heating from the stored heat content in the oceans—this has become known as "warming in the pipeline". This is the climate problem! A potential manifestation of this problem, in terms of the previously discussed issues of justice and equity is this: when the combined industrial activities of the developed world serve to eventually flood the entire country of Bangladesh—where can they turn for compensation and environmental justice?

A series of data figures below show the following (data compiled from various NASA and NOAA sources):

Since about 1995 warming in the Arctic regions have started to increase sharply relative to warming amounts at other latitudes.	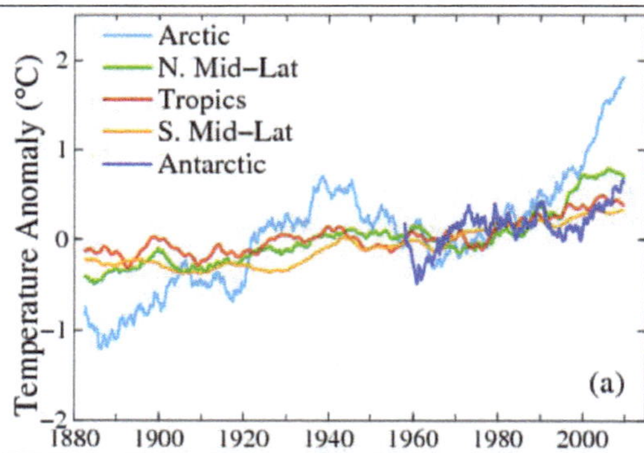
Data indicate a trend of about 6 C per century rise based on the 1980-2010 period. However, in recent years Arctic warming has substantially increased above that forecast.	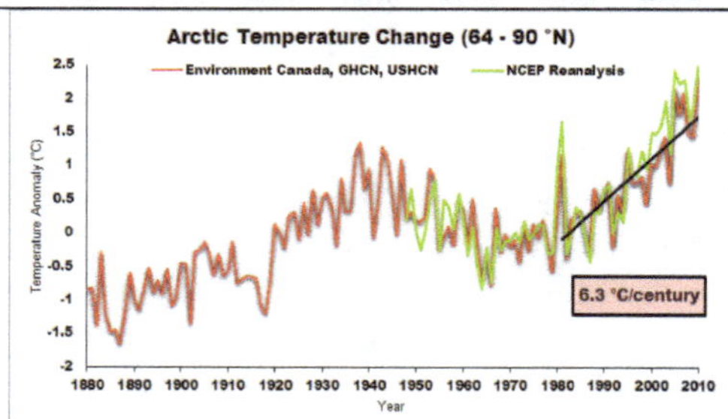

The most recent data shows an astounding amount of warming, up to 20 C larger than the North Polar Regions should be. This has the overall effect of accelerating the rate of Arctic Sea Ice loss and 2016 turns out to have set another record for the minimum sea ice extent in the Arctic Ocean. (See below). Whether this is a transient anomaly of the emergence of Arctic warming at a rate much higher than 6 C per century remains to be seen.

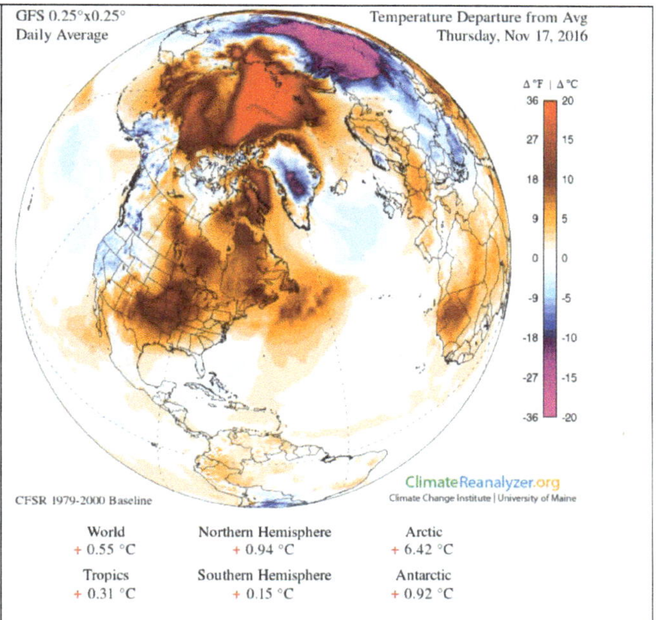

The 2016 Arctic sea ice melt season eclipsed the 2012 record for maximum sea ice loss and the amount of ice loss is significantly outside the normal range as established by the 1981-2010 baseline. This rapid rate of sea ice loss will lead eventually to an Arctic ocean which is free of ice for sometimes of year. Once predicted to be far away, it seems like this ice free period will be upon us within 10 years

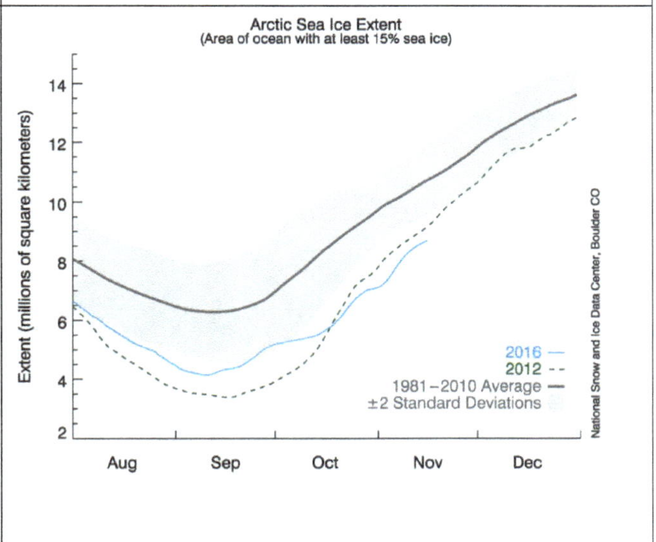

<table>
<tr>
<td>Permafrost contains a significant amount of frozen methane as the last ice age occur sufficiently fast that many "swamps" at Northerly latitudes were frozen quite fast. Arctic warming means the melting of the permafrost and the release of that methane, which is 35 times more potent, per molecule as CO_2. This data through Oct 2016 clearly shows that methane concentration in the atmosphere continues to increase, likely a direct result of this thawing permafrost.</td>
<td>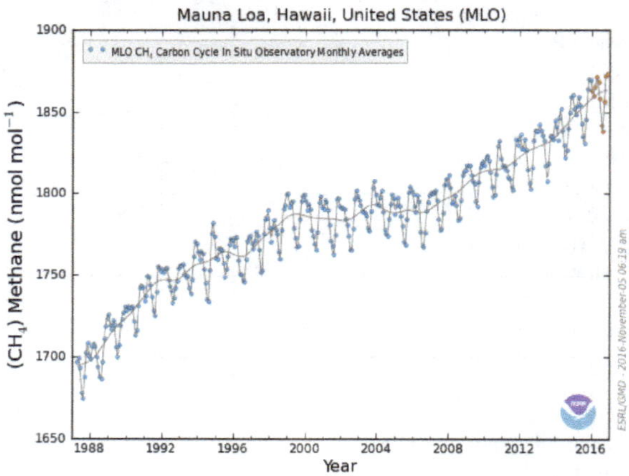
</td>
</tr>
<tr>
<td>It seems fitting to close this section in the spirit of the cartoon on the right. An ice free Arctic ocean will make family cruises to the North Pole possible in September. So future generations of parents can look forward to this trip much like currently taking their children to Disney world. However, when the parents and their children arrive to the liquid North Pole, the children are then allowed to as "How come you guys melted Santa's home?"</td>
<td>
</td>
</tr>
</table>

7 Concluding Remarks

It should be clear that we truly live in a world of finite resources that we are using at an alarming rate. While the chant of sustainability abounds, in the real world each of us are living in the most unstainable time in history. The end result of this business as usual trajectory does lie in accelerated climate change as well as accelerated resource depletion. No improved technology with more efficient resource extraction will overcome the basic problem that our current value system is skewed too much towards economic growth and too little toward the more fundamental values of environmental justice, social justice, equity and dignity for all the citizens of the Earth. Our collective value system and decision making needs to move away from purely economic considerations and towards a moral viewpoint that includes the overall cost of our choices. The current argument involving climate change as a potential violation of the public trust doctrine is gaining considerable momentum and legal standing (see the May 2016 NY times column by Erwin Chemerinsky). The public trust doctrine asserts that the governments holds natural resources "in trust" for future generations. The quality of our atmosphere and oceans are examples of these kinds of natural resources that are to be held "in trust". Clearly a value system centered on honoring the public trust, at the expense of potential economic development is required.

But why is a change in value system away from consumption and towards sustainability so difficult. To conclude we delve into this dilemma by using a well-known rubric in the business world for why the business culture cannot change. The 7 stated reasons that the business culture is resistant to change are the following:

1. There isn't any real need for the change
2. The change is going to make it harder for them to meet their needs
3. The risks seem to outweigh the benefits
4. They don't think they have the ability to make the change
5. They believe the change will fail
6. Change process is being handled improperly by management
7. The change is inconsistent with their values

Let's now briefly map these 7 reasons onto various kinds of quick reactions against changing our cultural consumption habits (see also the student exercise in Appendix 2):

1. **There isn't any real need for change** maps to: There is no evidence that suggests our current consumption habits are damaging in any way; consumption produces a better standard of living—why change that?
2. **The change is going to make it harder for them to meet their needs** maps to: Reducing material and energy consumption will significantly compromise my current lifestyle. Furthermore, since there is no evidence that compels me to make such a change, why would I/we do it?

3. **The risks seem to outweigh the benefits** maps to: My short term economic security is far more important than any long-term benefit for the planetary ecosystem.
4. **They don't think they have the ability to make the change** maps to I am an individual, what can I do that will actually make any impact?
5. **They believe the change will fail** maps to: Since there is no evidence that changing our consumption habits will have any positive effect then any such mandate to change will surely fail and have significant negative consequences.
6. **Change process is being handled improperly by management** maps to: We don't trust our government to make a fair set of regulations. We don't trust scientific advisors to the government to be unbiased. All policy recommendations serve only self-interests.
7. **The change is inconsistent with their values** maps to: Of course, it is! We are not part of nature; we control nature; nature does not control us. We are not in partnership with nature. The perfect, precise, logical, and ordered view of the Universe, as formulated by the mechanical philosophy is correct and Humans are Special.

Well that certainly is a daunting list but it does seem to be a succinct expression of the basic obstacles that we, as individuals, communities or nations face. To this author, as previously expressed, principle number 7 is key: we must change our value system. And while there is more than sufficient data and information presented in this chapter to make the reader remain pessimistic, that is a counterproductive position to take. As an individual interested in motivating sustainability you should remain optimistic and proactive and take heart in the following:

- Change can occur when consumers are properly informed. A lot of consumer decisions are made out of ignorance, not malice.
- Technological solutions exist to make significant impact if deployed now—particularly in the areas of renewable energy and nanotechnology.
- Consume Less—as an exercise, research the average American consumption of some product and choose to consume less than that average of the product. Then pass that tactic onto your social media peer group. That has impact.
- Humans are probably not **terminally stupid**—we have currently abandoned wisdom mostly for the sake of expediency and convenience. But we can still recover the wisdom needed to change our value system away from identity based consumption to a more holistic view of the world in which all of its systems have equal right to exist and flourish.

In a very real sense, collective humanity has to re-discover the sense of the sacred. For example, an aboriginal culture would be unlikely to dig through soil to tap an oil deposit because the dirt is sacred to their culture; the bones of their ancestors are buried there. In the US culture, one can just own land and do whatever you want with it—there is no cultural imperative to share its resources with others or to protect those resources. Indeed, this very ancient view of the world is exactly the view that needs to be re-acquired for humanity to become more sustainable.

Appendix 1: The University of Oregon Zero Waste Program

The concept behind zero waste is a good example of taking a whole system approach to consumption and in that process minimizing the amount or waste that is produced. Zero waste includes far more than just recycling. At its core, the zero waste movement emphasis reduced consumption through consumer choice of products that can be re-used. In turn, this movement can put pressure on the commercial sector to supply more re-usable goods to the consumer. Indeed, the power of the consumer could be large in this regard if conscientious consumers begin to refuse to buy products that cannot be reused.

A very good example of the kind of quantitative sustainability which can be achieved on a local scale through the zero waste concepts is the very successful Zero Waste program at the University of Oregon, which has now received national prominence. The mechanism of a solid waste audit becomes the principle manner which programs like this can actually, quantitatively, prove sustainability on the scale of their campus waste. Further mechanisms of waste to composting instead of landfills go a long way in helping to ensure zero waste.

The University of Oregon program is extensively detailed in this resource http://zerowaste.uoregon.edu/Book/ and also offers a toolkit for other campuses to use if they want to design and implement their own program (http://zerowaste.uoregon.edu/PDFdocuments/ZeroWasteToolkit.pdf). The specific case of a solid waste audit is discussed here: http://zerowaste.uoregon.edu/Book/#Chapter7—and this is an excellent source for student reading and experimental pilot projects. In order to effectively manage consumer waste, you must know what comprises it, in quantitative detail.

Since the implementation of this program depends almost exclusively on student labor (sometimes volunteer) the program becomes a visible manifestation of what sustainability can look like. Programs like this are real, they are quantitative, and they make an impact in the real world. All campuses are strongly encouraged to consider following the lead that University of Oregon has established.

Appendix 2: Some Student Exercises

Below we give some example student assignments based on some of the material presented in this chapter.

It is possible that the require links to databases in some of these assignments will not appear in the paper text—therefore, please go to http://homework.uoregon.edu/pub/class/springer.html to fully access these assignment suggestions.

Q1: Of the 7 stated reasons for resistance to change, write an editorial style 600-word piece on which 2 of these 7 reasons are the biggest obstacles to overcome and identify what strategy you would employ to overcome them.

Q2: Research the University of Oregon Zero Waste program and write a 600-word letter to the editor of your student newspaper advocating the benefits to your campus if it were to adopt this kind of program.

Q3: This data exercise makes use of *This Database. (Links to an external site.)*

An editorial in your student newspaper claims that US foreign policy has been oriented towards protecting US interests in the Middle East for the last 30 years because "almost all of our imported oil comes from the Middle East—everyone knows this".

Using the database above:

(a) Average the decade of the 80s, 90s, 00s and 2011–2015 of Annual import (units of thousand barrels a day) for two groups of countries:

Group A: Canada, Mexico and Venezuela
Group B: Saudi Arabia, Iraq, Kuwait

(b) Using this *This tool*—plot the data you just got and make a screen shot of that and insert it into your answer document.
(c) How does the Group B to Group A ratio change as a function of decade? What is that ratio *currently? (Links to an external site.)*
(d) Write a 150 word rebuttal letter to the Editor of your student newspaper

Q4: William Cronon's, *The Trouble With Wilderness (Links to an external site.),* (1996) is one of the most insightful works of environmental philosophy ever produced.

One of the fundamental tenets of environmentalism is the holiness of wilderness. It is considered a pure, pristine environment, "an island in the polluted sea of urban-industrial modernity," a landscape untouched by humanity. This concept is very much a human construct, however, and it is merely the latest version of an evolving human relationship to the wild.

The above is an excerpt from a review of Cronon's essay (linked above).

For this assignment, you are to read this essay and write a 600–800-word synopsis that included the following:

(a) Explain why this essay is insightful and how it serves to illuminate a kind of environmental hypocrisy
(b) In what ways does Cronon assert that the environmental movement is broken?
(c) How does Cronon's view of wilderness help society realize sustainability?

Q5: *Tesla's Powerwall Solution (Links to an external site.)*—let's do the energy math.

Here are the raw data/technical specifications for this new product needed to solve all parts of this question. The process is far more important than any numerical answers.

- Powerwall modular Storage = 10 KWHs.
- Continuous discharge power = 2KW.
- Battery Technology = Li-Ion @ 100 W hours per kg.

- Purchase + installation cost = \$4000.
- Residential electricity rate = 0.1\$ (10 cents) per KWH.
- Typical residential rooftop PV array rated at 4 KW maximum output (=noon on a sunny day).
- Average output in a day = 1/3 of peak.
- Rooftop cost + installation = \$1.80 per peak watt.
- Average number of suitably sunny days per year = 160.

Show all work involved in making the following estimates:

(a) What is the weight of the 10 KWH modular storage unit and how many hours does it take to discharge?
(b) What are the upfront costs to the homeowner for the purchase and installation of the storage unit and the PV rooftop?
(c) Averaged over the year, what is the average daily available solar energy that can be used to charge up the Powerwall system?
(d) How many hours, given this average daily available solar power, would it take to fully charge the 10 KWH system?
(e) Given your answer for d—explain what the homeowner would need to do to overcome this basic problem and estimate how much more this might cost.
(f) Given (d) and (e) and the residential electricity rate described above, what is the approximate payback time for just one of these storage modules?
(g) *The Global Lithium Market (Links to an external site.) (Links to an external site.)*—Refer to the table in that document that shows 36,000 metric tons of Lithium were available for products in the year 2014. Assume that 1/3 of world lithium production can be used to build Powerwall storage units for American houses. In the year 2017, world availability is projected to increase by 20% from the year 2014. Using this information, how many houses in America could have a Powerwall installed in the year 2017? What fraction of houses in America does this represent?
(h) Given what you know about scalability, write a **200-word** statement on whether or not the Powerwall is a scalable alternative energy solution for American houses.

Q6: Research and read about the Tesla GigaFactory and its planned production of 500,000 vehicles annually. Your team is hired as the consulting team for the GigaFactory. There are numerous studies on the lithium supply chain/problem, etc. that you can easily find via a Google Search. You are to access a variety of reports and synthesize a **500-word** consulting report to Tesla on the probability that they can actually meet this production goal with existing resources. If they cannot meet this goal, what needs to be then done to find new resources.

Q7: In this simulation, we explore the relationship between percentage of an ecosystem that is made toxic by some industrial process and the ability for a migrating species in that ecosystem to survive. The image is a snapshot of the simulation which shows an ecosystem in which 10% of it is made toxic—this is represented by the oil derrick symbols and the other 10%, the spider symbols, represent a migratory species

that uses the resources available on any of the green squares. The simulation was actually built to described the situation in Eastern Ecuador due to the highly exploitative oil exploration that occurred there in the 1980s and 1990s.

The goal here is to understand critical phenomena as there is a threshold of toxic ecosystem filling factor as a function of species migratory habits and timescales. In the control panel the user sets the initial population size, the total habit area and the breeding delay. The breeding delay is the number of pixels (green squares) of movement required for offspring. If one of the spiders moves randomly into one of the oil derrick squares, then it dies.

The links to the relevant simulations are here:

- http://homework.uoregon.edu/run/spiders/tests/test3.html (3% filling factor)
- http://homework.uoregon.edu/run/spiders/tests/test5.html (5% filling factor)
- http://homework.uoregon.edu/run/spiders/tests/test10.html (10% filling factor)
- http://homework.uoregon.edu/run/spiders/tests/test16.html (16% filling factor)

The goal of the student exercise is to determine the parameters for a population to be stable after 20 generations or so (there is a generation counter) and to explore the stability requirements in terms of initial population and breeding delay for each of the filling factor scenarios. After running a few of these the student should understand that once the filling factor gets too high, there is no way to produce a stable population.

Q8: Below are 5 quotes related to Consumerism on the part of some 20th century thinkers. I include them initially as anonymous so that you want be pre-biased. These quotes somewhat reinforce my closing remarks regarding our facile sacrificing of the sacred for the sake of the expedient and convenient (e.g. the 1950s throwaway society—see Dunning 1992).

In an essay of 1200 words (reserve 500 words for part C):

(a) Of the five quotes, select the one that best describes out situation and defend that selection.
(b) Integrating all of the quotes together, what is the combined message to us consumers?
(c) Your generation is really the last one that can make a difference. You are giving the graduation speech at your ceremony. In 500 words, what do you say to your peers that will change their value system with respect to consumerism?

The quotes:

1. *Armaments, universal debt, and planned obsolescence—those are the three pillars of Western prosperity. If war, waste, and moneylenders were abolished, you'd collapse. And while you people are over-consuming the rest of the world sinks more and more deeply into chronic disaster.*
2. *But even in the much-publicized rebellion of the young against the materialism of the affluent society, the consumer mentality is too often still intact: the standards of behavior are still those of kind and quantity, the security sought is still the security of numbers, and the chief motive is still the consumer's anxiety that he is missing out on what is "in". In this state of total consumerism—which is to say a state of helpless dependence on things and services and ideas and motives that we have forgotten how to provide ourselves—all meaningful contact between ourselves and the earth is broken. We do not understand the earth in terms either of what it offers us or of what it requires of us, and I think it is the rule that people inevitably destroy what they do not understand.*
3. *This is the postmodern desert inhabited by people who are, in effect, consuming themselves in the form of images and abstractions through which their desires, sense of identity, and memories are replicated and then sold back to them as products.*
4. *We seldom consider how much of our lives we must render in return for some object we barely want, seldom need, buy only because it was put before us...And this is understandable given the workings of our system where without a job we perish, where if we don't want a job and are happy to get by we are labeled irresponsible, non-contributing leeches on society. But if we hire a fleet of bulldozers, tear up half the countryside and build some monstrous factory, casino or mall, we are called entrepreneurs, job-creators, stalwarts of the community. Maybe we should all be shut away on some planet for the insane. Then again, maybe that is where we are.*
5. *On the way from the Renaissance to our days we have enriched our experience, but we have lost the concept of a Supreme Complete Entity which used to restrain our passions and our irresponsibility. We have placed too much hope in*

political and social reforms, only to find out that we were being deprived of our most precious possession: our spiritual life. In the East, it is destroyed by the dealings and machinations of the ruling party. In the West, commercial interests tend to suffocate it. This is the real crisis.

References

Agyeman, J., Bullard, R., & Evans, B. (2002). Exploring the nexus: bringing together sustainability, environmental justice and equity. *Space and Polity, 6*(1), 70–90.

Agyeman, J., Bullard, R., & Evans, B. (Eds.). (2003). *Just sustainabilities: Development in an unequal world*. Cambridge, MA: MIT Press.

Agyeman, J. (2005). *Sustainable communities and the challenge of environmental justice*. New York: New York University Press.

Armstrong, C. (2011) Global resource distribution. *Encyclopedia of Global Justice*, 441.

Babiak, M., & Boolish, M. (2009). Primary battery collection and recycling: A manufacture's perspective. In *13th International Battery Materials Recycling Seminar & Exhibit*, Ft. Lauderdale FL March 16–18 2009.

Bourke, S., & Meppem, T. (2000). Privileged narratives and fictions of consent in environmental discourse. *Local Environment, 5*(3), 299–310.

Bohanon, C. (2009). Economic recovery: Lessons from the Post-World War II period Mercatus Center on Policy, George Mason University. Retrieved from https://www.mercatus.org/system/files/PostWWII_Recovery_Bohanon_MOP112-%281%29-copy.pdf.

Brundtland Report. (1987). Report of the World Commission on Environmental and Development: Our Common Future. Retrieved from http://www.un-documents.net/our-common-future.pdf.

Chemerinsky, E. (2016). *New York Times Column May*. Retrieved from http://www.nytimes.com/roomfordebate/2016/05/23/can-citizens-sue-the-government-over-climate-change/citizens-have-a-right-to-sue-for-climate-change-action.

Cohen, L. (2004). A consumers' republic: The politics of mass consumption in postwar America. *Journal of Consumer Research, 31,* 1.

Connelly, S. (2007). Mapping sustainable development as a contested concept. *Local Environment, 12*(3), 259–278.

Chung, D., Elgqvist, E., & Santhanagopalan, S. (2015). Automotive Lithium-ion battery supply chain and US competiveness considerations. Retrieved from https://energy.gov/sites/prod/files/2015/06/f23/Lithium-ion%20Battery%20CEMAC.pdf.

Cronon, W. (1996). The trouble with wilderness: Or, getting back to the wrong nature, *Enviromental History* 1.1. Retrieved from http://www.williamcronon.net/writing/Trouble_with_Wilderness_Main.html.

Descartes, R. (1635). Discourse on method. Retrieved from https://plato.stanford.edu/entries/descartes/.

Dunning, A. (1992). How much is enough? The Consumer Society and the Future of the Earth The Worldwatch Environmental Alert Series.

Ernst, W. (2002). Global equity and sustainable earth resource consumption requires super-efficient extraction, conservation, recycling and ubiquitous. *Inexpensive Energy International Geology Review, 44,* 1072.

Gaines, L. (2014). The future of automotive lithium-ion battery recycling: Charting a sustainable course. *Sustainable Materials and Technologies, 1,* 2.

Gunder, M. (2006). Sustainability: Planning's saving grace or road to perdition? *Journal of Planning Education and Research, 26,* 208–221.

Howarth, R., & Marino, R. (2006). Nitrogen as the limiting nutrient for eutrophication in coastal marine ecosystems: Evolving views over three decades. *Limnology and Oceanography, 51,* 364.

Jacobs, M. (1999). Sustainable development: A contested concept. In A. Dobson (Ed.), *Fairness and Futurity. Essays on Environmental Sustainability and Social Justice*. Oxford: Oxford University Press.

Larrain, S., Leroy, J. P., & Nansen, K. (Eds). (2003). Citizen Contribution to the construction of sustainable societies. http://www.chilesustentable.net/web/wp-content/plugins/downloads-manager/upload/Comercio_Conosur-ingles.pdf.

Lebow, V. (1955, Spring). *Journal of Retailing*. Retrieved from http://www.gcafh.org/edlab/Lebow.pdf.

Living Planet Report. (2016). http://awsassets.panda.org/downloads/lpr_living_planet_report_2016.pdf.

Meadows, D., Meadows, D., Rangers, J., & Behrens, W. (1972). The Limits of Growth; commissioned by the Club of Rome.

Middleton, N., & O'Keefe, P. (2001). *Redefining sustainable development*. London: Pluto Press.

Mumford, L. (1934). *Book: Technics and civilizations.*

Mumford, L. (1938). *Book: The culture of cities.*

Nur, A. (1997). The collapse of ancient societies by great earthquakes. In *The 2nd SIS Cambridge Conference on Natural Catastrophes during Bronze Age Civilisations.*

Paris. (2015). United Nations framework convention on climate change. Retrieved from https://unfccc.int/resource/docs/2015/cop21/eng/l09r01.pdf.

Pianta, M., & Marchetti, R. (2007). The global justice movements: The transnational dimension. In D. della Porta (Ed.), *The global justice movement: A cross-national and transnational perspective* (pp. 29–51). Boulder CO: Paradigm.

Ren, J., Qin, D., Kang, S., Pu, J., & Jing, Z. (2004). Glacier variations and climate warming and drying in the central Himalayas. *Chinese Science Bulletin, 49,* 1.

Rockström, J., et al. (2009). Planetary boundaries: exploring the safe operating space for humanity. *Ecology and Society, 14*(2), 32.

Ryan, W., & Pittman, W. (1999). *Noah's flood: The new scientific discoveries about the event that changed history*. NY: Simon and Schuster.

Schlosberg, D. (1999). *Environmental justice and the new pluralism. The challenge of difference for environmentalism*. Oxford: Oxford University Press.

Schlosberg, D., & Collins, L. (2014). From environmental to climate justice: climate change and the discourse of environmental justice. *Wiley Interdisciplinary Reviews: Climate Change, 5,* 291.

Sen, A. (2009). *The idea of justice*. London: Allen Lane.

Shiva, V. (2015). *Soil not oil: Environmental justice in an age of climate crisis*. North Atlantic Books.

Springer, D. (2011). *Advertising in the Age of Persuasion: Building Brand America 1941–1961.* USA: Palgrave Macmillan.

Trenberth, K., et al. (2007). Estimates of the global water budget and its annual cycle using observational and model data. *Journal of Hydrometeorology, 8,* 758.

Tittensor, D., et al. (2014). A midterm analysis of progress towards international biodiversity targets. *Science, 346,* 6206.

Warlenius, R., Pierce, G., & Ramasar, V. (2015). The concept of "ecological debt" and its value for environmental justice. *Global Environmental Change, 30,* 21.

Wilkinson, R., & Pickett, K. (2009). *The spirit level: Why equality is better for everyone*. London: Allen Lane.

Worster, D. (1993). The shaky ground of sustainable development. In W. Sachs, (Ed.), *Global ecology: A new arena of political conflict* (pp. 115–131). Halifax, NS: Fernwood Books Ltd.

Wyman, R. (2013). The effects of population on the depletion of fresh water. *Population and Development Review, 39,* 687.

Index

J.P. Davim (ed.), *Curricula for Sustainability in Higher Education*,
Management and Industrial Engineering, DOI 10.1007/978-3-319-56505-7

FSC
www.fsc.org
MIX
Papier | Fördert
gute Waldnutzung
FSC® C083411

Kolibri 360 GmbH
Ferdinand-Jühlke-Straße 7
99095 Erfurt, Deutschland
produktsicherheit@kolibri360.de